NATEF Standards Job Sheets

Heating and Air Conditioning (A7)

Fourth Edition

Jack Erjavec
Ken Pickerill

CENGAGE Learning

Australia • Brazil • Japan • Korea • Mexico • Singapore • Spain • United Kingdom • United States

CENGAGE Learning

NATEF Standards Job Sheets
Heating and Air Conditioning (A7)
Fourth Edition
Jack Erjavec & Ken Pickerill

VP, General Manager, Skills and Planning: Dawn Gerrain

Director, Development, Global Product Management, Skills: Marah Bellegarde

Product Manager: Erin Brennan

Senior Product Development Manager: Larry Main

Senior Content Developer: Meaghan Tomaso

Product Assistant: Maria Garguilo

Marketing Manager: Linda Kuper

Market Development Manager: Jonathon Sheehan

Senior Production Director: Wendy Troeger

Production Manager: Mark Bernard

Content Project Management: S4Carlisle

Art Direction: S4Carlisle

Media Developer: Debbie Bordeaux

Cover image(s): © Snaprender/Dreamstime.com
© Shutterstock.com/gameanna
© i3alda/www.fotosearch.com
© IStockPhoto.com/tarras79
© IStockPhoto.com/tarras79
© Laurie Entringer
© IStockPhoto.com/Zakai
© stoyanh/www.fotosearch.com

Library of Congress Control Number: 2014931167

ISBN-13: 978-1-1116-4703-2
ISBN-10: 1-111-64703-8

Cengage Learning
200 First Stamford Place, 4th Floor
Stamford, CT 06902
USA

Cengage Learning is a leading provider of customized learning solutions with office locations around the globe, including Singapore, the United Kingdom, Australia, Mexico, Brazil, and Japan. Locate your local office at:
www.cengage.com/global

Cengage Learning products are represented in Canada by Nelson Education, Ltd.

To learn more about Cengage Learning, visit **www.cengage.com**
Purchase any of our products at your local college store or at our preferred online store **www.cengagebrain.com**

Printed in the United States of America
1 2 3 4 5 6 7 18 17 16 15 14

CONTENTS

NATEF TASK LIST FOR HEATING AND AIR CONDITIONING

Required Supplemental Tasks (RST)

Shop and Personal Safety

1. Identify general shop safety rules and procedures.
2. Utilize safe procedures for handling of tools and equipment.
3. Identify and use proper placement of floor jacks and jack stands.
4. Identify and use proper procedures for safe lift operation.
5. Utilize proper ventilation procedures for working within the lab/shop area.
6. Identify marked safety areas.
7. Identify the location and the types of fire extinguishers and other fire safety equipment; demonstrate knowledge of the procedures for using fire extinguishers and other fire safety equipment.
8. Identify the location and use of eyewash stations.
9. Identify the location of the posted evacuation routes.
10. Comply with the required use of safety glasses, ear protection, gloves, and shoes during lab/shop activities.
11. Identify and wear appropriate clothing for lab/shop activities.
12. Secure hair and jewelry for lab/shop activities.
13. Demonstrate awareness of the safety aspects of supplemental restraint systems (SRS), electronic brake control systems, and hybrid vehicle high-voltage circuits.
14. Demonstrate awareness of the safety aspects of high-voltage circuits (such as high intensity discharge (HID) lamps, ignition systems, injection systems, etc.).
15. Locate and demonstrate knowledge of material safety data sheets (MSDS).

Tools and Equipment

1. Identify tools and their usage in automotive applications.
2. Identify standard and metric designation.
3. Demonstrate safe handling and use of appropriate tools.
4. Demonstrate proper cleaning, storage, and maintenance of tools and equipment.
5. Demonstrate proper use of precision measuring tools (i.e., micrometer, dial-indicator, dial-caliper).

Preparing Vehicle for Service

1. Identify information needed and the service requested on a repair order.
2. Identify purpose and demonstrate proper use of fender covers, mats.
3. Demonstrate use of the three C's (concern, cause, and correction).
4. Review vehicle service history.
5. Complete work order to include customer information, vehicle identifying information, customer concern, related service history, cause, and correction.

Preparing Vehicle for Customer

1. Ensure vehicle is prepared to return to customer per school/company policy (floor mats, steering wheel cover, etc.).

Maintenance and Light Repair (MLR) Task List

A. General

A.1.	Research applicable vehicle and service information, vehicle service history, service precautions, and technical service bulletins.	Priority Rating 1

B. Refrigeration System Components

B.1.	Inspect and replace A/C compressor drive belts, pulleys, and tensioners; determine necessary action.	Priority Rating 1
B.2.	Identify hybrid vehicle A/C system electrical circuits and the service/safety precautions.	Priority Rating 2
B.3.	Inspect A/C condenser for airflow restrictions; determine necessary action.	Priority Rating 1

C. Heating, Ventilation, and Engine Cooling Systems Diagnosis and Repair

C.1.	Inspect engine cooling and heater systems hoses; perform necessary action.	Priority Rating 1

D. Operating Systems and Related Controls Diagnosis and Repair

D.1.	Inspect A/C-heater ducts, doors, hoses, cabin filters, and outlets; perform necessary action.	Priority Rating 1
D.2.	Identify the source of A/C system odors.	Priority Rating 2

Automobile Service Technology (AST) Task List

A. General: A/C System Diagnosis and Repair

A.1.	Identify and interpret heating and air conditioning problems; determine necessary action.	Priority Rating 1
A.2.	Research applicable vehicle and service information, vehicle service history, service precautions, and technical service bulletins.	Priority Rating 1
A.3.	Performance test A/C system; identify problems.	Priority Rating 1
A.4.	Identify abnormal operating noises in the A/C system; determine necessary action.	Priority Rating 2
A.5.	Identify refrigerant type; select and connect proper gauge set; record temperature and pressure readings.	Priority Rating 1
A.6.	Leak test A/C system; determine necessary action.	Priority Rating 1
A.7.	Inspect condition of refrigerant oil removed from A/C system; determine necessary action.	Priority Rating 2

A.8.	Determine recommended oil and oil capacity for system application.	Priority Rating 1
A.9.	Using a scan tool, observe and record related HVAC data and trouble codes.	Priority Rating 3

B. Refrigeration System Component Diagnosis and Repair

B.1.	Inspect and replace A/C compressor drive belts, pulleys, and tensioners; determine necessary action.	Priority Rating 1
B.2.	Inspect, test, service or replace A/C compressor clutch components and/or assembly; check compressor clutch air gap; adjust as needed.	Priority Rating 2
B.3.	Remove, inspect, and reinstall A/C compressor and mountings; determine recommended oil quantity.	Priority Rating 2
B.4.	Identify hybrid vehicle A/C system electrical circuits and service/safety precautions.	Priority Rating 2
B.5.	Determine need for an additional A/C system filter; perform necessary action.	Priority Rating 3
B.6.	Remove and inspect A/C system mufflers, hoses, lines, fittings, O-rings, seals, and service valves; perform necessary action.	Priority Rating 2
B.7.	Inspect A/C condenser for airflow restrictions; perform necessary action.	Priority Rating 1
B.8.	Remove, inspect, and reinstall receiver/drier or accumulator/drier; determine recommended oil quantity.	Priority Rating 2
B.9.	Remove, inspect, and install expansion valve or orifice (expansion) tube.	Priority Rating 1
B.10.	Inspect evaporator housing water drain; perform necessary action.	Priority Rating 1
B.11.	Diagnose A/C system conditions that cause the protection devices (pressure, thermal, and PCM) to interrupt system operation; determine necessary action.	Priority Rating 2
B.12.	Determine procedure to remove and reinstall evaporator; determine required oil quantity.	Priority Rating 2
B.13.	Remove, inspect, and reinstall condenser; determine required oil quantity.	Priority Rating 2

C. Heating, Ventilation, and Engine Cooling Systems Diagnosis and Repair

C.1.	Inspect engine cooling and heater systems hoses; perform necessary action.	Priority Rating 1
C.2.	Inspect and test heater control valve(s); perform necessary action.	Priority Rating 2
C.3.	Diagnose temperature control problems in the heater/ventilation system; determine PCM to interrupt system operation; determine necessary action.	Priority Rating 2
C.4.	Determine procedure to remove, inspect, and reinstall heater core.	Priority Rating 2

D. Operating Systems and Related Controls Diagnosis and Repair

D.1.	Inspect and test A/C-heater blower motors, resistors, switches, relays, wiring, and protection devices; perform necessary action.	Priority Rating 1
D.2.	Diagnose A/C compressor clutch control systems; determine necessary action.	Priority Rating 2
D.3.	Diagnose malfunctions in the vacuum, mechanical, and electrical components and controls of the heating, ventilation, and A/C (HVAC) system; determine necessary action.	Priority Rating 2
D.4.	Inspect and test A/C-heater control panel assembly; determine necessary action.	Priority Rating 3
D.5.	Inspect and test A/C-heater control cables, motors, and linkages; perform necessary action.	Priority Rating 3
D.6.	Inspect A/C-heater ducts, doors, hoses, cabin filters, and outlets; perform necessary action.	Priority Rating 1

D.7. Identify the source of A/C system odors. Priority Rating 2

D.8. Check operation of automatic or semi-automatic heating, ventilation, and air-conditioning (HVAC) control systems; determine necessary action. Priority Rating 2

E. Refrigerant Recovery, Recycling, and Handling

E.1. Perform correct use and maintenance of refrigerant handling equipment according to equipment manufacturer's standards. Priority Rating 1

E.2. Identify and recover A/C system refrigerant. Priority Rating 1

E.3. Recycle, label, and store refrigerant. Priority Rating 1

E.4. Evacuate and charge A/C system; add refrigerant oil as required. Priority Rating 1

MASTER AUTOMOBILE SERVICE TECHNOLOGY (MAST) TASK LIST

A. General: A/C System Diagnosis and Repair

A.1. Identify and interpret heating and air conditioning problems; determine necessary action. Priority Rating 1

A.2. Research applicable vehicle and service information, vehicle service history, service precautions, and technical service bulletins. Priority Rating 1

A.3. Performance test A/C system; identify problems. Priority Rating 1

A.4. Identify abnormal operating noises in the A/C system; determine necessary action. Priority Rating 2

A.5. Identify refrigerant type; select and connect proper gauge set; record temperature and pressure readings. Priority Rating 1

A.6. Leak test A/C system; determine necessary action. Priority Rating 1

A.7. Inspect condition of refrigerant oil removed from A/C system; determine necessary action. Priority Rating 2

A.8. Determine recommended oil and oil capacity for system application. Priority Rating 1

A.9. Using a scan tool, observe and record related HVAC data and trouble codes. Priority Rating 3

B. Refrigeration System Component Diagnosis and Repair

B.1. Inspect and replace A/C compressor drive belts, pulleys, and tensioners; determine necessary action. Priority Rating 1

B.2. Inspect, test, service or replace A/C compressor clutch components and/or assembly; check compressor clutch air gap; adjust as needed. Priority Rating 2

B.3. Remove, inspect, and reinstall A/C compressor and mountings; determine recommended oil quantity. Priority Rating 2

B.4. Identify hybrid vehicle A/C system electrical circuits and service/safety precautions. Priority Rating 2

B.5. Determine need for an additional A/C system filter; perform necessary action. Priority Rating 3

B.6. Remove and inspect A/C system mufflers, hoses, lines, fittings, O-rings, seals, and service valves; perform necessary action. Priority Rating 2

B.7. Inspect A/C condenser for airflow restrictions; perform necessary action. Priority Rating 1

B.8. Remove, inspect, and reinstall receiver/drier or accumulator/drier; determine recommended oil quantity. Priority Rating 2

B.9. Remove, inspect, and install expansion valve or orifice (expansion) tube. Priority Rating 1

B.10. Inspect evaporator housing water drain; perform necessary action. Priority Rating 1

B.11. Diagnose A/C system conditions that cause the protection devices (pressure, thermal, and PCM) to interrupt system operation; determine necessary action. Priority Rating 2

B.12.	Determine procedure to remove and reinstall evaporator; determine required oil quantity.	Priority Rating 2
B.13.	Remove, inspect, and reinstall condenser; determine required oil quantity.	Priority Rating 2

C. Heating, Ventilation, and Engine Cooling Systems Diagnosis and Repair

C.1.	Inspect engine cooling and heater systems hoses; perform necessary action.	Priority Rating 1
C.2.	Inspect and test heater control valve(s); perform necessary action.	Priority Rating 2
C.3.	Diagnose temperature control problems in the heater/ventilation system; determine PCM to interrupt system operation; determine necessary action.	Priority Rating 2
C.4.	Determine procedure to remove, inspect, and reinstall heater core.	Priority Rating 2

D. Operating Systems and Related Controls Diagnosis and Repair

D.1.	Inspect and test A/C-heater blower motors, resistors, switches, relays, wiring, and protection devices; perform necessary action.	Priority Rating 1
D.2.	Diagnose A/C compressor clutch control systems; determine necessary action.	Priority Rating 2
D.3.	Diagnose malfunctions in the vacuum, mechanical, and electrical components and controls of the heating, ventilation, and A/C (HVAC) system; determine necessary action.	Priority Rating 2
D.4.	Inspect and test A/C-heater control panel assembly; determine necessary action.	Priority Rating 3
D.5.	Inspect and test A/C-heater control cables, motors, and linkages; perform necessary action.	Priority Rating 3
D.6.	Inspect A/C-heater ducts, doors, hoses, cabin filters, and outlets; perform necessary action.	Priority Rating 1
D.7.	Identify the source of A/C system odors.	Priority Rating 2
D.8.	Check operation of automatic or semi-automatic heating, ventilation, and air-conditioning (HVAC) control systems; determine necessary action.	Priority Rating 2

E. Refrigerant Recovery, Recycling, and Handling

E.1.	Perform correct use and maintenance of refrigerant handling equipment according to equipment manufacturer's standards.	Priority Rating 1
E.2.	Identify and recover A/C system refrigerant.	Priority Rating 1
E.3.	Recycle, label, and store refrigerant.	Priority Rating 1
E.4.	Evacuate and charge A/C system; add refrigerant oil as required.	Priority Rating 1

DEFINITION OF TERMS USED IN THE TASK LIST

To clarify the intent of these tasks, NATEF has defined some of the terms used in the task listings. For a good understanding of what the task includes, refer to this glossary while reading the task list.

add	To increase fluid or pressure to the correct level or amount.
adjust	To bring components to specified operational settings.
analyze	To examine the relationship of components of an operation. Assess the condition of a component or system.
assemble (reassemble)	To fit together the components of a device.
bleed	To remove air from a closed system.
CAN	Acronym for *Controller Area Network.* CAN is a network protocol (SAE J2284/ISO 15765-4) used to interconnect a network of electronic control modules.

charge	To bring to a specified state (e.g., battery or air conditioning system).
check	To verify condition by performing an operational or comparative examination.
clean	To rid components of foreign matter for the purpose of reconditioning, repairing, measuring, and reassembling.
confirm	To acknowledge something has happened with firm assurance.
demonstrate	To show or exhibit the knowledge of a theory or procedure.
determine	To establish the procedure to be used to perform the necessary repair.
determine necessary action	Indicates that the diagnostic routine or routines is the primary emphasis of a task. The student is required to perform the diagnostic steps and communicate the diagnostic outcomes and corrective actions required to address the concern or problem. The training program determines the communication method (worksheet, test, verbal communication, or other means deemed appropriate) and whether the corrective procedures for these tasks are actually performed.
diagnose	To identify the cause of a problem.
differentiate	To perceive the difference in or between one thing to other things.
disassemble	To separate a component's parts in preparation for cleaning, inspection, or service.
discharge	To empty a storage device or system.
drain	To use gravity to empty a container.
evacuate	To remove air, fluid, or vapor from a closed system by the use of a vacuum pump.
fill (refill)	To bring fluid level to a specified point or volume.
flush	To internally clean a component or system.
high voltage	Voltages of 50 volts and higher.
identify	To establish the identity of a vehicle or component prior to service; to determine the nature or degree of a problem.
install (reinstall)	To place a component in its proper position in a system.
leak test	To locate the source of leaks in a component or system.
locate	Determine or establish a specific spot or area.
maintain	To keep something at a specified level, position, rate, etc.
measure	To compare and determine existing dimensions/values to specified dimensions through the use of calibrated instruments and gauges.
network	A system of interconnected electrical modules or devices.
on-board diagnostics (OBD)	Diagnostic protocol that monitors computer inputs and outputs for failures.
perform	To accomplish a procedure in accordance with established methods and standards.
perform necessary action	Indicates that the student is to perform the diagnostic routine(s) and perform the corrective action item. Where various scenarios (conditions or situations) are presented in a single task, at least one of the scenarios must be accomplished.
pressure test	To use air or fluid pressure to determine the integrity, condition, or operation of a component or system.
priority ratings	Indicates the minimum percentage of tasks, by area, that a program must include in its curriculum in order to be certified in that area.
purge	To remove air or fluid from a closed system.

remove	To disconnect and separate a component from a system.
replace	To exchange an unserviceable component with a new or rebuilt component; to reinstall a component.
research	To make a thorough investigation into a situation or matter.
select	To choose the correct part or setting during assembly or adjustment.
service	To perform a procedure as specified in the owner's manual or service manual.
test	To verify a condition through the use of meters, gauges, or instruments.
verify	To confirm that a problem exists after hearing the customer's concern; or, to confirm the effectiveness of a repair.

CROSS-REFERENCE GUIDES

RST Task	**Job Sheet**
Shop and Personal Safety	
1.	1
2.	2
3.	2
4.	2
5.	2
6.	2
7.	3
8.	1
9.	2
10.	1
11.	1
12.	1
13.	4, 5
14.	6
15.	7
Tools and Equipment	
1.	8
2.	8
3.	8
4.	8
5.	8
Preparing Vehicle for Service	
1.	9
2.	9
3.	9
4.	9
5.	9
Preparing Vehicle for Customer	
1.	9

MLR Task	**Job Sheet**
A.1	11
B.1	18
B.2	21
B.3	24
C.1	29

MLR Task	Job Sheet
D.1	37
D.2	38

AST Task	Job Sheet
A.1	10
A.2	11
A.3	12
A.4	13
A.5	14
A.6	15
A.7	16
A.8	16
A.9	17
B.1	18
B.2	19
B.3	20
B.4	21
B.5	22
B.6	23
B.7	24
B.8	23
B.9	23
B.10	25
B.11	27
C.1	29
C.2	30
C.3	32
D.1	33
D.2	34
D.3	35
D.4	36
D.5	35
D.6	37
D.7	38
D.8	39
E.1	40
E.2	41
E.3	42
E.4	43

MAST Task	Job Sheet
A.1	10
A.2	11
A.3	12
A.4	13
A.5	14
A.6	15
A.7	16
A.8	16
A.9	17
B.1	18
B.2	19
B.3	20

MAST Task	Job Sheet
B.4	21
B.5	22
B.6	23
B.7	24
B.8	23
B.9	23
B.10	25
B.11	26
B.12	27
B.13	28
C.1	29
C.2	30
C.3	31
C.4	32
D.1	33
D.2	34
D.3	35
D.4	36
D.5	35
D.6	37
D.7	38
D.8	39
E.1	40
E.2	41
E.3	42
E.4	43

PREFACE

The automotive service industry continues to change with the technological changes made by automobile, tool, and equipment manufacturers. Today's automotive technician must have a thorough knowledge of automotive systems and components, good computer skills, exceptional communication skills, good reasoning, the ability to read and follow instructions, and above average mechanical aptitude and manual dexterity.

This new edition, like the last, was designed to give students a chance to develop the same skills and gain the same knowledge that today's successful technician has. This edition also reflects the changes in the guidelines established by the National Automotive Technicians Education Foundation (NATEF) in 2013.

The purpose of NATEF is to evaluate technician training programs against standards developed by the automotive industry and recommend qualifying programs for certification (accreditation) by ASE (National Institute for Automotive Service Excellence). Programs can earn ASE certification upon the recommendation of NATEF. NATEF's national standards reflect the skills that students must master. ASE certification through NATEF evaluation ensures that certified training programs meet or exceed industry-recognized, uniform standards of excellence.

At the expense of much time and many minds, NATEF has assembled a list of basic tasks for each of their certification areas. These tasks identify the basic skills and knowledge levels that competent technicians have. The tasks also identify what is required for a student to start a successful career as a technician.

In June 2013, after many discussions with the industry, NATEF established a new model for automobile program standards. This new model is reflected in this edition and covers the new standards, which are based on three (3) levels: Maintenance & Light Repair (MLR), Automobile Service Technician (AST), and Master Automobile Service Technician (MAST). Each successive level includes all the tasks of the previous level in addition to new tasks. In other words, the AST task list includes all of the MLR tasks, plus additional tasks. The MAST task list includes all of the AST tasks, plus additional tasks specifically for MAST.

Most of the content in this book are job sheets. These job sheets relate to the tasks specified by NATEF, according to the appropriate certification level. The main considerations during the creation of these job sheets were student learning and program certification by NATEF. Students are guided through standard industry-accepted procedures. While they are progressing, they are asked to report their findings, as well as offer their thoughts on the steps they have just completed. The questions asked of the students are thought provoking and require students to apply what they know to what they observe.

The job sheets were also designed to be generic. That is, whenever possible, the tasks can be performed on any vehicle from any manufacturer. Also, completion of the sheets does not require the use of specific brands of tools and equipment. Rather, students use what is available. In addition, the job sheets can be used as a supplement to any good textbook.

Also included are description and basic use of the tools and equipment listed in NATEF's standards. The standards recognize that not all programs have the same needs, nor do all programs teach all of the NATEF tasks. Therefore, the basic philosophy for the tools and equipment requirement is that the training should be as thorough as possible with the tools and equipment necessary for those tasks.

Theory instruction and hands-on experience of the basic tasks provide initial training for employment in automotive service or further training in any or all of the specialty areas. Competency in the tasks indicates to employers that you are skilled in that area. You need to know the appropriate theory, safety, and support information for each required task. This should include identification and use of the required tools, testing, and measurement equipment required for the tasks, the use of current reference and training materials, the proper way to write work orders and warranty reports, and the storage, handling, and use of hazardous materials as required by the "Right to Know" law, and federal, state, and local governments.

Words to the Instructor: We suggest you grade these job sheets based on completion and reasoning. Make sure the students answer all questions. Then, look at their reasoning to see if the task was actually completed and to get a feel for their understanding of the topic. It will be easy for students to copy others' measurements and findings, but each student should have their own base of understanding and that will be reflected in their explanations.

Words to the Student: While completing the job sheets, you have a chance to develop the skills you need to be successful. When asked for your thoughts or opinions, think about what you observed. Think about what could have caused those results or conditions. You are not being asked to give accurate explanations for everything you do or observe. You are only asked to think. Thinking leads to understanding. Good technicians are good because they have a basic understanding of what they are doing and why they are doing it.

Jack Erjavec and Ken Pickerill

HEATING AND AIR CONDITIONING SYSTEMS

To prepare you to learn what you should learn from completing the job sheets, some basics must be covered. This discussion begins with an overview of heating and air conditioning systems. Emphasis is placed on what they do and how they work, including the major components of heating and air conditioning systems and their role in the efficient operation of heating and air conditioning systems of all designs.

Preparation for work on an automobile would not be complete without addressing certain safety issues. This discussion covers what you should and should not do while working on heating and air conditioning systems, including the proper ways to deal with hazardous and toxic materials.

NATEF's task list for Heating and Air Conditioning certification is given, with definitions of some terms used to describe the tasks. This list gives you a good look at what the experts say you need to know before you can be considered competent to work on heating and air conditioning systems.

Following the task list are descriptions of the various tools and types of equipment you must be familiar with. These are the tools you will use to complete the job sheets. They are also the tools NATEF has identified as necessary for servicing heating and air conditioning systems.

After the tool discussion is a cross-reference guide that shows which NATEF tasks are related to specific job sheets. In most cases, there are single job sheets for each task. Some tasks are part of a procedure, in which case one job sheet may cover two or more tasks. The remainder of the book contains the job sheets.

HEATING SYSTEMS

The main components of an automotive heating system are the heater core, the heater control valve, the blower motor and the fan, and the heater and defroster ducts. The heating system works with the engine's cooling system and converts the heat from the coolant circulating inside the engine to hot air, which is blown into the passenger compartment. A heater hose transfers hot coolant from the engine to the heater control valve and from there to the heater core inlet. As the coolant circulates through the core, heat is transferred from the coolant to the tubes and fins of the core. Air blown through the core by the blower motor and fan then picks up heat from the surfaces of the core and transfers it into the passenger compartment. After giving up its heat, the coolant is then pumped out through the heater core outlet, where it is returned to the engine's cooling system to be heated again.

The heater control valve (also called the water flow valve) controls the flow of coolant into the heater core from the engine. In a closed position, the valve allows no flow of hot coolant to the heater core, thereby keeping the heater core cool. In an open position, the valve allows heated coolant to circulate through the heater core, thereby maximizing heater efficiency.

Cable-operated heater control valves are controlled directly by the heater control lever on the dashboard. Thermostatically controlled valves have a liquid-filled capillary tube located in the discharge air stream off the heater core. This tube senses air temperature, and the valve modulates the flow of water to maintain a constant temperature, regardless of engine speed or temperature.

Most heater control valves are vacuum operated. When a vacuum signal reaches the valve, a diaphragm inside the valve is raised, either opening or closing the valve against an opposing spring. When the temperature selection on the heater control panel is changed, vacuum to the valve is vented and the valve returns to its original position. Some vehicles don't use a heater control valve; instead, a heater door controls how much heat is released into the passenger compartment from the heater core.

The blower motor is usually located in the heater housing assembly. A multiposition switch on the control panel controls its speed. The switch works in connection with a stepped resistor block normally located near the heater housing. The resistor block controls blower motor speed. The typical resistor block has three or four resistors in series with the blower motor. The multiposition switch directs voltage to different locations in this series circuit. When low fan speed is selected, voltage is present at the beginning of the series circuit. High resistance causes slower operation of the blower motor. When high speed is selected, voltage is present at the end, or near the end, of the resistor assembly. Low resistance allows for higher fan speeds. On some vehicles, when the engine is running, the blower motor runs constantly at low speed.

Transferring heated air from the heater core to the passenger compartment is the job of the heater and defroster ducts. The ducts are typically part of a large plastic shell that connects to the necessary inside and outside vents. This ductwork also has mounting points for the evaporator and heater core assemblies. Also contained inside the duct are the doors required to direct air to the floor, dash, and/or windshield. Sometimes the duct is connected directly to the vents; other times, hoses are used.

Ventilation System

The ventilation system is designed to supply outside air to the passenger compartment through upper or lower vents, or both. Several systems are used to vent air into the passenger compartment. The most common is the flow-through system. In this arrangement, a supply of outside air, which is called ram air, flows into the car when it is moving. When the car is not moving, the heater fan can produce a steady flow of outside air. In operation, air is circulated throughout the passenger compartment. From there the air is forced outside the vehicle through an exhaust area.

AIR CONDITIONING SYSTEMS

It is important to know the basic theory of automotive air conditioning systems. Understanding how an A/C system moves heat from the confined space of the passenger compartment and dissipates it into the atmosphere will help you diagnose A/C problems.

An air conditioning system (Figure 1) is designed to pump heat from one point to another. All materials or substances, even those as cold as –459° Fahrenheit, have heat in them. Also, heat always flows from a warmer object to a colder one. The greater the temperature difference between the objects, the greater the amount of heat flow.

Objects can be in one of three forms or states: solid, liquid, or gas. When objects change from one state to another, large amounts of heat can be transferred. For example, when water temperature goes below 32°F, water changes from a liquid to a solid (ice). If the temperature of water is raised to 212°F, the liquid turns into a gas (steam). An interesting thing occurs when water, or any matter, changes from a solid to a liquid and then from a liquid to a gas. Additional heat is necessary to change the state of the substance, even though this heat does not register on a thermometer. For example, ice at 32°F requires heat to change into water, which will also be at 32°F. Additional heat raises the temperature of the water until it reaches the boiling point of 212°F. More heat is required to change water into steam. If the temperature of the steam were measured, however, it would also be 212°F. The amount of heat necessary to change the state of a substance is called latent heat—or hidden heat—because it cannot be measured with a thermometer. This hidden heat is the basic principle behind all air conditioning systems.

Pressure on a substance, such as a liquid, changes its boiling point. The greater the pressure on a liquid, the higher the boiling point. If pressure is placed on a vapor, the vapor condenses at a higher-than-normal temperature. In addition, as the pressure on a substance is reduced, the boiling point can also be reduced. For example,

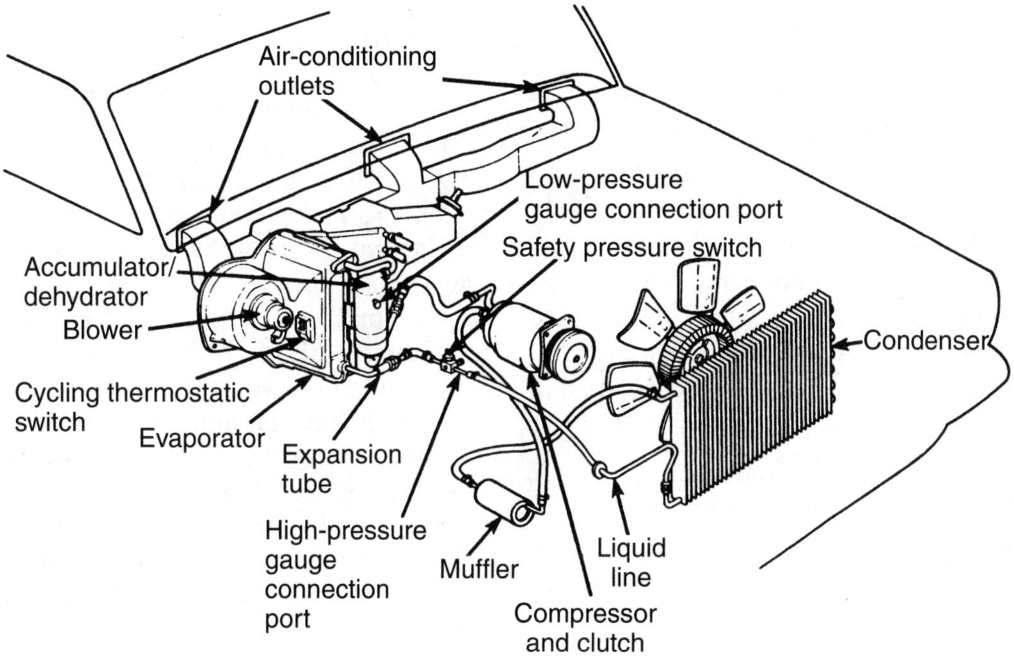

Figure 1 A typical air conditioning system.

the boiling point of water is 212°F. Increasing the pressure on the fluid can increase the boiling point. Reducing the pressure or placing the fluid in a vacuum can decrease it.

Refrigerants

The substance used to remove heat from the inside of an air-conditioned vehicle is called the refrigerant.

To replace in A/C systems, the manufacturers have agreed to use R-134a. Since about 1993, R-134a has been used as a refrigerant in automotive air conditioning systems. This refrigerant may also be referred to as SUVA. R-134a is a hydrofluorocarbon (HFC).

R-134a has the capability of contributing to the greenhouse effect when released into the air. Therefore, the recovery and recycling of R-134a is mandatory by law. Laws have also been passed that dictate the certification of equipment and technicians. A new refrigerant has been introduced to supersede R-134a called HFO-1234yf. This new refrigerant will require a different set of gauge and dedicated recovery and recycling tools. R-134 has fallen from favor for its greenhouse gas potential. As of this time, the EPA does not mandate the use of HFO-1234yf, but most automakers plan to switch to the new refrigerant by around 2018. In Europe the new refrigerant was mandatory in 2014.

Basic Operation

Refrigerants are used to carry heat from the inside of the vehicle to outside the vehicle. Automotive refrigerants have a low boiling point (the point at which evaporation occurs). As a refrigerant changes state, it absorbs a large amount of heat. Since the heat that it absorbs is from the inside of the vehicle, passengers become cooler.

To understand how a refrigerant is used to cool the interior of a vehicle, the effects of pressure and temperature on it must be understood. If the pressure of the refrigerant is high, so is its temperature. If the pressure is low, so is its temperature. Therefore, changing its pressure can change the temperature of a refrigerant.

To absorb heat, the temperature and pressure of the refrigerant is kept low. To dissipate heat, the temperature and pressure is kept high. As the refrigerant absorbs heat, it changes from a liquid to a vapor. As it dissipates heat, it changes from a vapor to a liquid. The change from a vapor to a liquid is called condensation. These two changes of state—evaporation and condensation—occur continuously as the refrigerant circulates through the air conditioning system.

An A/C system is a closed, pressurized system. It consists of a compressor, condenser, receiver/drier or accumulator, expansion valve or orifice tube, and an evaporator. In a basic air conditioning system, the heat is absorbed and transferred via the following steps.

1. Refrigerant leaves the compressor as a high-pressure, high-temperature vapor.
2. By removing heat via the condenser, the vapor becomes a high-pressure, high-temperature liquid.
3. Moisture and contaminants are removed by the receiver/drier, where the cleaned refrigerant is stored until it is needed.
4. The expansion valve controls the flow of refrigerant into the evaporator.
5. Heat is absorbed from the air inside the passenger compartment by the low-pressure, refrigerant enters the evaporator as a refrigerant, causing the liquid to vaporize and greatly decrease its temperature.
6. The refrigerant returns to the compressor as a low-pressure, low-temperature vapor.

The best way to understand the purpose of an A/C system's major components is to divide the systems into two sides: the high side and the low side. The high side is under high pressure and high temperature. The low side is the low-pressure, low-temperature side of the system.

The Compressor

The compressor separates the high-pressure and low-pressure sides of the system. Its primary purpose is to draw the low-pressure and low-temperature vapor from the evaporator and compress this vapor into high-temperature, high-pressure vapor. This action results in the refrigerant having a higher temperature than surrounding air and enables the condenser to condense the vapor back to a liquid. The secondary purpose of the compressor is to circulate or pump the refrigerant through the A/C system under the different pressures required for proper operation. The compressor is located on the engine and is driven by the engine's crankshaft via a drive belt.

Although many types of compressors are in use today, they are usually one of three types. A *piston compressor* can have its pistons arranged in an in-line, axial, radial, or V design. It is designed to have an intake stroke and a compression stroke for each cylinder. On the intake stroke, the refrigerant from the low side of the system (evaporator) is drawn into the compressor. Refrigerant is taken in through intake reed valves, which are one-way valves that control the flow of refrigerant vapors into the cylinder. During the compression stroke, the vaporous refrigerant is compressed. This increases both the pressure and the temperature of the heat-carrying refrigerant. The outlet or discharge-side reed valves then open to allow the refrigerant to move to the condenser. The outlet reed valves are the beginning of the high side of the system.

A common variation of a piston-type compressor is the variable displacement compressor. These compressors not only act as a regular compressor, but they also control the amount of refrigerant that passes through the evaporator. The pistons are connected to a wobble-plate. The angle of the wobble-plate determines the stroke of the pistons and is controlled by the difference in pressure between the compressor's outlet and inlet. When the stroke of the pistons increases, more refrigerant is being pumped and there is increased cooling.

A *rotary vane compressor* doesn't have pistons. Instead, it consists of a rotor with several vanes and a carefully shaped housing. As the compressor shaft rotates, the vanes and housing form chambers. The refrigerant is drawn through the suction port into these chambers, which become smaller as the rotor turns. The discharge port is located at the point where the gas is completely compressed.

The *scroll-type compressor* has a movable scroll and a fixed scroll that provide an eccentric-like motion. As the compressor's crankshaft rotates, the movable scroll forces the refrigerant against the fixed scroll and toward the center of the compressor. This motion pressurizes the refrigerant. The pressure of air moving in a circular pattern increases as it moves toward the center of the circle. A delivery port is positioned at the center of the compressor and allows the high-pressure refrigerant to flow into the air conditioning system.

Refrigerant Oils

Normally the only source of lubrication for a compressor is the oil mixed with the refrigerant. Because of the loads and speeds at which the compressor operates, proper lubrication is a must for long compressor life. R-134a systems require a synthetic oil, poly alkaline glycol (PAG). There are a number of different blends of PAG oil; always use the one recommended by the vehicle manufacturer or compressor manufacturer.

Compressor Clutches

Compressors are equipped with an electromagnetic clutch as part of the compressor pulley assembly. The clutch is designed to engage the pulley to the compressor shaft when the clutch coil is energized. The clutch provides a way for turning the compressor on or off.

The clutch is driven by power from the engine's crankshaft. When the clutch is engaged, power is transmitted from the pulley to the compressor shaft by the clutch drive plate. When the clutch is not engaged, the compressor shaft does not rotate, and the pulley freewheels.

The clutch is engaged by a magnetic field and disengaged by springs when the magnetic field is broken. When the controls call for compressor operation, the electrical circuit to the clutch is completed, the magnetic clutch is energized, and the clutch engages the compressor. When the electrical circuit is opened, the clutch disengages the compressor.

Condenser

The condenser consists of coiled refrigerant tubing mounted in a series of thin cooling fins to provide maximum heat transfer in a minimum amount of space. The condenser is normally mounted just in front of the vehicle's radiator. It receives the full flow of ram air from the movement of the vehicle or airflow from the radiator fan when the vehicle is standing still.

The condenser condenses or liquefies the high-pressure, high-temperature vapor coming from the compressor. To do so, it must give up its heat. The condenser receives very hot (normally 200° to 400°F), high-pressure refrigerant vapor from the compressor through its discharge hose. The refrigerant vapor enters the inlet at the top of the condenser, and as the hot vapor passes down through the condenser coils, heat moves from the hot refrigerant into the cooler air as it flows across the condenser coils and fins. This process causes a large quantity of heat to be transferred to the outside air and the refrigerant to change from a high-pressure hot vapor to a high-pressure warm liquid. This high-pressure warm liquid flows from the outlet at the bottom of the condenser through a line to the receiver/drier, or to the refrigerant metering device if an accumulator instead of a drier is used.

Receiver/Drier

The receiver/drier is a storage tank for the liquid refrigerant from the condenser. The refrigerant flows into the upper portion of the receiver tank, which contains a bag of desiccant (moisture-absorbing material, such as silica alumina or silica gel). The desiccant absorbs unwanted water and moisture in the refrigerant. As the refrigerant flows through the outlet at the lower portion of the receiver, a mesh screen filters it.

Included in many receiver/driers are a high-pressure fitting, a pressure relief valve, and a sight glass for determining the state and condition of the refrigerant in the system.

Accumulator

Most late-model systems are equipped with an accumulator rather than a receiver/drier. The accumulator is connected into the low side, at the outlet of the evaporator. The accumulator also contains a desiccant and is designed to store excess refrigerant and to filter and dry the refrigerant. If liquid refrigerant flows out of the evaporator, it will be collected by and stored in the accumulator. The main purpose of an accumulator is to prevent liquid from entering the compressor.

Thermostatic Expansion Valve or Orifice Tube

The refrigerant flow to the evaporator must be controlled to obtain maximum cooling, while ensuring complete evaporation of the liquid refrigerant within the evaporator. This is the job of a thermostatic expansion valve (TEV or TXV) or a fixed orifice tube.

The TEV is mounted at the inlet to the evaporator and separates the high-pressure side of the system from the low-pressure side. The TEV regulates refrigerant flow to the evaporator to prevent evaporator flooding or starving. In operation, the TEV regulates the refrigerant flow to the evaporator by balancing the inlet flow to the outlet temperature.

Both externally and internally equalized TEVs are used in air conditioning systems. The only difference between the two valves is that the external TEV uses an equalizer line connected to the evaporator outlet line as a means of sensing evaporator outlet pressure. The internal TEV senses evaporator inlet pressure through an internal equalizer passage. Both valves have a capillary tube to sense evaporator outlet temperature.

Like the TEV, the orifice tube is the dividing point between the high- and low-pressure parts of the system. However, its metering or flow rate control does not depend on comparing evaporator pressure and temperature. It is a fixed orifice. The flow rate is determined by pressure difference across the orifice and by subcooling. Subcooling is additional cooling of the refrigerant in the bottom of the condenser after it has changed from vapor to liquid. The flow rate through the orifice is more sensitive to subcooling than to pressure difference.

Evaporator

The evaporator, like the condenser, consists of a refrigerant coil mounted in a series of thin cooling fins. It provides a maximum amount of heat transfer in a minimum amount of space. The evaporator is usually located beneath the dashboard or instrument panel.

On receiving the low-pressure, low-temperature liquid refrigerant from the thermostatic expansion valve or orifice tube in the form of an atomized (or droplet) spray, the evaporator serves as a boiler or vaporizer. The regulated flow of refrigerant boils immediately. Heat from the core surface is lost to the boiling and vaporizing refrigerant, which is cooler than the core, thereby cooling the core. The air passing over the evaporator loses its heat to the cooler surface of the core, thereby cooling the air inside the car. As the process of heat loss from air to the evaporator core surface is taking place, any moisture (humidity) in the air condenses on the outside of the evaporator core and is drained off as water. This dehumidification of air adds to passenger comfort. It can also be used to control the fogging of windows.

Through the metering, or controlling, action of the thermostatic expansion valve or orifice tube, greater or lesser amounts of refrigerant are provided in the evaporator to adequately cool the car under all heat load conditions. If too much refrigerant is allowed to enter, the evaporator floods. This results in poor cooling due to the higher pressure (and temperature) of the refrigerant. The refrigerant can neither boil away rapidly nor vaporize. On the other hand, if too little refrigerant is metered, the evaporator starves. Poor cooling again results because the refrigerant boils away or vaporizes too quickly before passing through the evaporator.

Refrigerant Lines

All the major components of the system have inlet and outlet connections that accommodate either flare or O-ring fittings. There are three major refrigerant lines. Suction lines are located between the outlet side of the evaporator and the inlet side or suction side of the compressor. They carry the low-pressure, low-temperature refrigerant vapor to the compressor, where it again is recycled through the system. Suction lines are always distinguished from the discharge lines by touch and size. They are cold to the touch. The suction line is larger in diameter than the liquid line because refrigerant in a vapor state takes up more room than refrigerant in a liquid state.

The discharge or high-pressure line connects the compressor to the condenser. The liquid lines connect the condenser to the receiver/drier and the receiver/drier to the inlet side of the expansion valve. Through these lines, the refrigerant travels in its path from a gas state (compressor outlet) to a liquid state (condenser outlet) and then to the inlet side of the expansion valve, where it vaporizes on entry to the evaporator. Discharge and liquid lines are always very warm to the touch and easily distinguishable from the suction lines.

Aluminum tubing is commonly used to connect air conditioning components where flexibility is not required. R-134a systems are required to be fitted with quick-disconnect fittings through the system. These also have hoses specially made for R-134a. They have an additional layer of rubber that serves as a barrier to prevent the refrigerant from escaping through the pores of the hose.

Blower Motor and Fan

The blower motor/fan assembly is located in the evaporator housing. The blower, which is basically the same type as those used in heater-systems, draws warm air from the passenger compartment, forces it over the coils and fins of the evaporator, and blows the cooled, cleaned, and dehumidified air into the passenger compartment. A fan switch controls the blower motor. During cold weather, the blower motor/fan assembly provides the airflow, and the heater core provides the heat for the passenger compartment.

Evaporator Pressure Control Systems

Evaporator controls maintain back pressure in the evaporator. Because of the refrigerant temperature-pressure relationship, the effect is to regulate evaporator temperature. The temperature is controlled to a point that provides effective air cooling but prevents the freezing of moisture that condenses on the evaporator.

In this type of system, the compressor operates continually when dash controls are in the air conditioning position. An evaporator pressure control valve automatically controls evaporator outlet air temperature. These types of valves throttle the flow of refrigerant out of the evaporator as required to establish a minimum evaporator pressure and thereby prevent freezing of condensation on the evaporator core.

Cycling Clutch Systems

In cycling clutch systems, the compressor is run intermittently by controlling the application and release of its clutch by a thermostatic or pressure switch. The thermostatic switch senses the evaporator's outlet air temperature through a capillary tube that is part of the switch assembly. With a high sensing temperature, the thermostatic switch is closed and the compressor clutch is energized. As the evaporator outlet temperature drops to a preset level, the thermostatic switch opens the circuit to the compressor clutch. The compressor then ceases to operate until such time as the evaporator temperature rises above the switch setting. The term cycling clutch derives from this on-and-off operation. In effect, the thermostatic switch is calibrated to allow the lowest possible evaporator outlet temperature that prevents the freezing of condensation on the evaporator.

Variations of the cycling clutch system include a system with a TEV and a system with an orifice tube.

Compressor Controls

There are many controls used to monitor and trigger the compressor during its operational cycle. Each of these represents the most common protective control devices designed to ensure safe and reliable operation of the compressor.

An ambient temperature switch senses outside air temperature and is designed to prevent compressor clutch engagement when air conditioning is not required or when compressor operation might cause internal damage to seals and other parts.

On some vehicles, the ambient switch is located in the air inlet duct of the air conditioning system or near the radiator. It is not required on systems with a thermostatic or pressure switch.

In cycling clutch systems, a thermostatic switch is placed in series with the compressor clutch circuit so it can turn the clutch on or off. It deenergizes the clutch and stops the compressor if the evaporator is at the freezing point. It also may control the air temperature by turning the compressor on and off intermittently.

When the temperature of the evaporator approaches the freezing point, the thermostatic switch opens the circuit and disengages the compressor clutch. The compressor remains inoperative until the evaporator temperature rises to the preset temperature, at which time the switch closes and compressor operation resumes.

A pressure cycling switch is connected in series with the compressor clutch. Like the thermostatic switch, the turning on and off of the pressure cycling switch controls the operation of the compressor.

The low-pressure cutoff or discharge pressure switch is located on the high side of the system and senses any low-pressure conditions. This switch is tied into the compressor clutch circuit, allowing it to immediately disengage the clutch when the pressure falls too low.

The high-pressure cutoff switch is normally located near the compressor and is also wired in series with the compressor clutch. This switch is designed to open and disengage the clutch at a specified high pressure. When the pressure drops below a specified pressure, it again closes and reengages the clutch.

A high-pressure relief valve may be installed on the receiver/drier, compressor, or elsewhere in the high side of the system. It is a high-pressure protection device that opens to bleed off any excessive pressure in the system.

The compressor control valve regulates the crankcase pressure in some compressors. It has a pressure-sensitive bellows assembly exposed to the suction side. This bellows acts on a ball-and-pin valve, which is exposed to high-side pressure. The bellows also controls a bleed port that is exposed to the low side. The control valve is continuously modulating, changing the displacement of the compressor according to pressure or temperature.

An electronic cycling clutch switch prevents evaporator freeze-up by sending a signal to the Powertrain Control Module (PCM). The PCM, in turn, cycles the compressor on and off by monitoring suction line temperature. If the temperature gets too low, the switch opens the input circuit to the computer, which causes the A/C clutch relay to open, which disengages the compressor clutch. Often this switch is the thermostatic switch at the evaporator.

Case/Duct Systems

The purpose of a heater/air conditioner/case and duct system is twofold. It is used to house the heater core and the evaporator and to direct the selected supply air through these components into the passenger compartment. The selected supply air can be either fresh (outside) or recirculated air, depending on the system mode. After the air is heated or cooled, it is delivered to the floor outlet, dash panel outlets, or the defrost outlets.

Manufacturers use two basic types of duct systems. The stacked core reheat system uses the heater control valve as the main control. All air enters the passenger compartment through the heater core. An access door controls only fresh or recirculated air. The blend air reheat system uses a mixture of heated and cooled air to provide comfort. The duct system brings in outside air and/or recirculated air across the heater core and evaporator. The door controls the source of the air entering into the passenger compartment.

Temperature Control Systems

Temperature control systems for air conditioners usually are connected with heater controls. Most heater and air conditioning systems use the same plenum chamber for air distribution. Two types of air conditioning controls are used: manual/semiautomatic and automatic.

Air conditioner manual/semiautomatic temperature controls operate in a manner similar to heater controls. Depending on the control setting, doors are opened and closed to direct airflow. The amount of cooling is controlled manually through the use of control settings and blower speed.

Automatic or electronic temperature control systems maintain a specific temperature automatically. To maintain a selected temperature, heat sensors send signals to a computer unit that controls compressor, heater valve, blower, and plenum door operation. A typical electronic control system might contain a coolant temperature sensor, in-car temperature sensor, outside temperature sensor, high-side temperature switch, low-side temperature switch, low-pressure switch, vehicle speed sensor, throttle position sensor, sunload sensor, and power steering cutout switch.

Three types of control panels may be found: manual, push-button, or touch pad. All serve the same purpose—they provide operator input control for the air conditioning and heating system. Some control panels have features that other panels do not have, such as provisions to display in-car and outside air temperature in degrees. Provisions are made on the control panel for operator selection of an in-car temperature between 65° (18°C) and 85°F (29°C) in one-degree increments. Some have an override feature that provides for a setting of either 60° (16°C) or 90°F (32°C). Either of these two settings overrides all in-car temperature control circuits to provide maximum cooling or heating conditions.

Multiplexing

Multiplexing is an in-vehicle networking system used to transfer data between electronic modules through a serial data bus. Serial data is electronically coded information that is transmitted by one computer and received and displayed by another computer. Serial data is information that is digitally coded and transmitted in a series of data bits. The data transmission rate is referred to as the baud rate. Baud rate refers to the number of data bits that can be transmitted in a second.

With multiplexing, fewer dedicated wires are required for each function, and this reduces the size of the wiring harness. Using a serial data bus reduces the number of wires by combining the signals on a single wire through time division multiplexing. Information is sent to individual control modules that control each function, such as engine controls, transmission controls, antilock braking, and many different accessories.

Multiplexing also eliminates the need for redundant sensors because the data from one sensor is available to all electronic modules. Multiplexing also allows for greater vehicle content flexibility because functions can be added through software changes, rather than adding another module or modifying an existing one.

The common multiplex system is called the CAN or Controller Area Network. It is used to interconnect a network of electronic control modules.

Hybrid Systems

A hybrid electric vehicle (HEV) uses one or more electric motors and an engine to propel the vehicle. Depending on the design of the system, the engine may move the vehicle by itself, assist the electric motor while it is moving the vehicle, or it may drive a generator to charge the vehicle's batteries. The electric motor may power the vehicle by itself or assist the engine while it is propelling the vehicle. Many hybrids rely exclusively on the electric motor(s) during slow speed operation, the engine at higher speeds, and both during some certain driving conditions. Complex electronic controls monitor the operation of the vehicle, based on the current operating conditions; electronics control the engine, electric motor, and generator.

A hybrid's electric motor is powered by high-voltage batteries, which are recharged by a generator driven by the engine and through regenerative braking. Regenerative braking is the process by which a vehicle's kinetic energy can be captured while it is decelerating and braking. The electric drive motors become generators driven by the vehicle's wheels. These generators take the kinetic energy, or the energy of the moving vehicle, and change it into energy that charges the batteries. The magnetic forces inside the generator cause the drive wheels to slow down. A conventional brake system brings the vehicle to a safe stop.

The engines used in hybrids are specially designed for the vehicle and electric assist. Therefore, they can operate more efficiently; resulting in very good fuel economy and very low tailpipe emissions. HEVs can provide the same performance, if not better, as a comparable vehicle equipped with a larger engine.

There are primarily two types of hybrids: the parallel and the series designs. A parallel hybrid electric vehicle uses either the electric motor or the gas engine to propel the vehicle, or both. The engine in a true series hybrid electric vehicle is used only to drive the generator that keeps the batteries charged. The vehicle is powered only by the electric motor(s). Most current HEVs are considered as having a series/parallel configuration because they have the features of both designs.

Although most current hybrids are focused on fuel economy, the same ideas can be used to create high-performance vehicles. Hybrid technology is also influencing off-the-road performance. By using individual motors at the front and rear drive axles, additional power can be applied to certain drive wheels, when needed.

SAFETY

In an automotive repair shop, there is great potential for serious accidents simply because of the nature of the business and the equipment used. Through carelessness, the automotive repair industry can be one of the most dangerous occupations. However, the chances of being injured while working on a car are close to nil if you learn to work safely and use common sense. Safety is the responsibility of everyone in the shop.

Personal Protection

Some procedures, such as grinding, result in tiny particles of metal and dust that are thrown off at very high speeds. These metal and dirt particles can easily get into your eyes, causing scratches or cuts on your eyeball. Pressurized gases and liquids escaping a ruptured or damaged hose fitting can spray a great distance. If these chemicals get into your eyes, they can cause blindness. This is especially true of refrigerants. Refrigerants can quickly freeze an eye and cause serious damage.

Eye protection should be worn whenever you are exposed to these risks. To be safe, you should wear safety glasses whenever you are working in the shop. Some procedures may require that you wear other eye protection in addition to safety glasses; for example, when cleaning parts with a pressurized spray, you should wear a face shield. The face shield not only gives added protection to your eyes but it also protects the rest of your face.

If chemicals such as battery acid, fuel, or solvents get into your eyes, flush them continuously with clean water. Have someone call a doctor and get medical help immediately.

Your clothing should be well fitted and comfortable but made with strong material. Loose, baggy clothing can easily get caught in moving parts and machinery. Some technicians prefer to wear coveralls or shop coats to protect their personal clothing. Your work clothing should offer you some protection but should not restrict your movement.

Long hair and loose, hanging jewelry can create the same type of hazard as loose-fitting clothing. They can get caught in moving engine parts and machinery. If you have long hair, tie it back or tuck it under a cap.

Never wear rings, watches, bracelets, or neck chains. These can easily get caught in moving parts and cause serious injury.

Always wear leather or similar material shoes or boots with non-slip soles. Steel-tipped safety shoes can give added protection to your feet. Jogging or basketball shoes, street shoes, or sandals are inappropriate in the shop.

Good hand protection is often overlooked. A scrape, cut, or burn can limit your effectiveness at work for many days. A well-fitted pair of heavy work gloves should be worn during operations such as grinding and welding or when handling high-temperature components. Always wear approved rubber gloves when handling strong and dangerous caustic chemicals.

When working on A/C systems, protect your hands from exposure to the refrigerant. Escaping refrigerant can cause frostbite to the area of skin exposed to the leak. In addition to wearing gloves, it is wise to use a rag when connecting and disconnecting service gauges to an A/C system.

Many technicians wear thin, surgical-type latex gloves whenever they are working on vehicles. These offer little protection against cuts but do offer protection against disease and grease buildup under and around your fingernails. These gloves are comfortable and are quite inexpensive.

Accidents can be prevented simply by the way you act. Following are some guidelines to follow while working in a shop. The list does not include everything you should or shouldn't do; it merely gives some things to think about.

- Never smoke while working on a vehicle or while working with any machine in the shop.
- Playing around is not fun when it sends someone to the hospital.
- To prevent serious burns, keep your skin away from hot metal parts, such as the radiator, exhaust manifold, tailpipe, catalytic converter, and muffler.
- Always disconnect electric engine cooling fans when working around the radiator. Many of these will turn on without warning and can easily chop off a finger or hand. Make sure you reconnect the fan after you have completed your repairs.
- When working with a hydraulic press, make sure the pressure is applied in a safe manner. It is generally wise to stand to the side when operating the press.
- Properly store all parts and tools by putting them away in a place where people will not trip over them. This practice not only cuts down on injuries, it also reduces time wasted looking for a misplaced part or tool.

Work Area Safety

Your entire work area should be kept clean and safe. Any oil, coolant, or grease on the floor can make it slippery. To clean up oil, use commercial oil absorbent. Keep all water off the floor. Water is slippery on smooth floors, and electricity flows well through water. Aisles and walkways should be kept clean and wide enough to move through easily. Make sure the work areas around machines are large enough to safely operate the machine.

Gasoline is a highly flammable volatile liquid. Something that is flammable catches fire and burns easily. A volatile liquid is one that vaporizes very quickly. Flammable volatile liquids are potential firebombs. Always keep gasoline or diesel fuel in an approved safety can, and never use gasoline to clean your hands or tools.

Handle all solvents (or any liquids) with care to avoid spillage. Keep all solvent containers closed, except when pouring. Proper ventilation is very important in areas where volatile solvents and chemicals are used. Solvent and other combustible materials must be stored in approved and designated storage cabinets or rooms with adequate ventilation. Never light matches or smoke near flammable solvents and chemicals, including battery acids.

Oily rags should also be stored in an approved metal container. When oily, greasy, or paint-soaked rags are left lying about or are not stored properly, they can spontaneously combust. Spontaneous combustion refers to fire that starts by itself, without a match.

Disconnecting the vehicle's battery before working on the electrical system, or before welding, can prevent fires caused by a vehicle's electrical system. To disconnect the battery, remove the negative or ground cable from the battery and position it away from the battery.

Know where all the shop's fire extinguishers are located. Fire extinguishers are clearly labeled as to type and types of fire they should be used on. Make sure you use the correct type of extinguisher for the type of fire you are dealing with. A multipurpose dry chemical fire extinguisher will put out ordinary combustibles, flammable liquids, and electrical fires. Never put water on a gasoline fire—the water will just spread the fire. The proper fire extinguisher smothers the flames.

During a fire, never open doors or windows unless it is absolutely necessary; the extra draft will only make the fire worse. Make sure the fire department is contacted before or during your attempt to extinguish a fire.

Working Safely on High-Voltage Systems

Hybrid vehicles have high-voltage systems. Most often, the A/C compressor is powered by high voltage. Careless handling of some components can lead to serious injury, including death. Always follow and adhere to the precautions given by the manufacturer. All service procedures should be followed exactly as defined by the manufacturer. Being careless and/or not following the procedures can cause serious injury and can cause the battery to explode!

Most high-voltage circuits are identifiable by size and color. The cables have thicker insulation and are typically colored orange. The connectors are also colored orange. On some vehicles, the high-voltage cables are enclosed in an orange shielding or casing, again the orange indicates high voltage. In addition, the high-voltage battery pack and most high-voltage components have "High Voltage" caution labels. Be careful not to touch these wires and parts.

Wear insulating gloves, commonly called lineman's gloves, when working on or around the high-voltage system. These gloves must be class "0" rubber insulating gloves, rated at 1000-volts. Also, to protect the integrity of the insulating gloves, as well as you, wear leather gloves over the insulating gloves while doing a service.

Make sure they have no tears, holes, or cracks and that they are dry. Electrons are very small and can enter through the smallest of holes in your gloves. The integrity of the gloves should be checked before using them. To check the condition of the gloves, blow enough air into each one so they balloon out. Then fold the open end over to seal the air in. Continue to slowly fold that end of the glove toward the fingers. This will compress the air. If the glove continues to balloon as the air is compressed, it has no leaks. If any air leaks out, the glove should be discarded. All gloves, new and old, should be checked before they are used.

There are other safety precautions that should always be adhered to when working on an electric drive vehicle:

- Obtain the necessary training before working on these vehicles.
- Be sure to perform each repair operation following the test procedures defined by the manufacturer.
- Disable or disconnect the high-voltage system before performing services to those systems. Do this according to the procedures given by the manufacturer.
- Anytime the engine is running in a hybrid vehicle, the generator is producing high voltage and care must be taken to prevent being shocked.
- Before doing any service to an electric drive vehicle, make sure the power to the electric motor is disconnected or disabled.
- Systems may have a high-voltage capacitor that must be discharged after the high-voltage system has been isolated. Make sure to wait the prescribed amount of time (normally about 10 minutes) before working on or around the high-voltage system.

- After removing a high-voltage cable, cover the terminal with vinyl electrical tape.
- Always use insulated tools.
- Alert other technicians that you are working on the high-voltage systems with a warning sign such as "High-Voltage Work: Do Not Touch."
- Many electric motors have a strong permanent magnet in them; individuals with a pacemaker should not handle these parts.
- When an electric drive vehicle needs to be towed into the shop for repairs, make sure it is not towed on its drive wheels. Doing this will drive the generator(s), which can overcharge the batteries and cause them to explode. Always tow these vehicles with the drive wheels off the ground or move them on a flat bed.

Tool and Equipment Safety

Careless use of simple hand tools, such as wrenches, screwdrivers, and hammers, causes many shop accidents that could be prevented. Keep all hand tools free of grease and in good condition. Tools that slip can cause cuts and bruises. If a tool slips and falls into a moving part, it can fly out and cause serious injury.

Use the proper tool for the job, and make sure the tool is of professional quality. Using poorly made tools or the wrong tools can damage parts, the tool itself, or you. Never use broken or damaged tools.

Safety around power tools is very important. Serious injury can result from carelessness. Always wear safety glasses when using power tools. If the tool is electrically powered, make sure it is properly grounded. Before using it, check the wiring for bare wires and for cracks in the insulation. When using electrical power tools, never stand on a wet or damp floor. Never leave a running power tool unattended.

When using compressed air, wear safety glasses or a face shield, or both. Particles of dirt and pieces of metal, blown by the high-pressure air, can penetrate your skin or get into your eyes.

Before using a compressed air tool, check all hose connections. Always hold an air nozzle or air control device securely when starting or shutting off the compressed air. A loose nozzle can whip suddenly and cause serious injury. Never point an air nozzle at anyone. Never use compressed air to blow dirt from your clothes or hair. Never use compressed air to clean the floor or workbench.

Always be careful when raising a vehicle on a lift or a hoist. Adapters and hoist plates must be positioned correctly to prevent damage to the vehicle's underbody. There are specific lift points that allow the weight of the vehicle to be evenly supported by the adapters or hoist plates. The correct lift points can be found in the vehicle's service manual. Before operating any lift or hoist, carefully read the operating manual and follow the operating instructions.

Once you know the lift supports are properly positioned under the vehicle, raise the lift until the supports contact the vehicle. Then, check the supports to make sure they are in full contact with the vehicle. Shake the vehicle to make sure it is securely balanced on the lift, and then raise the lift to the desired working height. Before working under a car, make sure the lift's locking devices are engaged.

A vehicle can be raised off the ground by a hydraulic jack. The jack's lifting pad must be positioned under an area of the vehicle's frame or at one of the manufacturer's recommended lift points. Never place the pad under the floor pan or under steering and suspension components, which are easily damaged by the weight of the vehicle. Always position the jack so the wheels of the vehicle can roll as the vehicle is being raised.

Safety stands, also called jack stands, should be placed under a sturdy chassis member, such as the frame or axle housing, to support the vehicle after it has been raised by a jack. Once the safety stands are in position, the hydraulic pressure in the jack should be slowly released until the weight of the vehicle is on the stands. Never move under a vehicle when it is supported only by a hydraulic jack. Rest the vehicle on the safety stands before moving under the vehicle.

Cleaning parts is a necessary step in most repair procedures. Always wear the appropriate protection when using chemical, abrasive, and thermal cleaners.

Vehicle Operation

When the customer brings a vehicle in for service, certain driving rules should be followed to ensure your safety and the safety of those working around you. For example, before moving a car into the shop, buckle your safety belt. Make sure no one is near, the way is clear, and there are no tools or parts under the car before you start the engine.

Check the brakes before putting the vehicle in gear. Then, drive slowly and carefully in and around the shop.

If the engine must be running while work is done on the car, block the wheels to prevent the car from moving. Place the transmission in park for automatic transmissions or in neutral for manual transmissions. Set the parking (emergency) brake. Never stand directly in front of or behind a running vehicle.

Run the engine only in a well-ventilated area to avoid the danger of poisonous carbon monoxide (CO) in the engine exhaust. CO is an odorless but deadly gas. Most shops have an exhaust ventilation system; always use it. Connect the hose from the vehicle's tailpipe to the intake for the vent system. Make sure the vent system is turned on before running the engine. If the work area does not have an exhaust venting system, use a hose to direct the exhaust out of the building.

HAZARDOUS MATERIALS AND WASTES

A typical shop contains many potential health hazards for those working in it. These hazards can cause injury, sickness, impairment, discomfort, and even death. Here is a short list of the different classes of hazards.

- Chemical hazards are caused by high concentrations of vapors, gases, or solids in the form of dust.
- Hazardous wastes are substances that are the result of a service.
- Physical hazards include excessive noise, vibration, pressures, and temperatures.
- Ergonomic hazards are conditions that impede normal or proper body position and motion.

There are many government agencies charged with ensuring safe work environments for all workers. These include the Occupational Safety and Health Administration (OSHA), Mine Safety and Health Administration (MSHA), and National Institute for Occupational Safety and Health (NIOSH). These agencies, in addition to state and local governments, have instituted regulations that must be understood and followed. Everyone in a shop is responsible for adhering to these regulations.

An important part of a safe work environment is the employees' knowledge of potential hazards. Right-to-know laws concerning all chemicals protect every employee in the shop. The general intent of right-to-know laws is for employers to provide their employees with a safe working place as it relates to hazardous materials.

All employees must be trained about their rights under the legislation, the nature of the hazardous chemicals in their workplace, and the contents of the labels on the chemicals. All information about each chemical must be posted on Material Safety Data Sheets (MSDS) and must be accessible. The manufacturer of the chemical must give these sheets to its customers if it is asked to do so. They detail the chemical composition and precautionary information for all products that can present a health or safety hazard.

Employees must become familiar with the general uses, protective equipment, accident or spill procedures, and any other information regarding the safe handling of the hazardous material. This training must be given to employees annually and provided to new employees as part of their job orientation.

All hazardous material must be properly labeled, indicating what health, fire, or reactivity hazard it poses and what protective equipment is necessary when handling each chemical. The manufacturer of the hazardous materials must provide all warnings and precautionary information, which must be read and understood by the user before use. A list of all hazardous materials used in the shop must be posted for the employees to see.

Shops must maintain documentation on the hazardous chemicals in the workplace, proof of training programs, records of accidents or spill incidents, and satisfaction of employee requests for specific chemical information via the MSDS. A general right-to-know compliance procedure manual must be used in the shop.

When handling any hazardous materials or hazardous waste, make sure you follow the required procedures for handling such material. Wear the proper safety equipment listed on the MSDS, which includes the use of approved respirator equipment.

Some of the common hazardous materials that automotive technicians use are cleaning chemicals, fuels (gasoline and diesel), paints and thinners, battery electrolyte (acid), used engine oil, refrigerants, and engine coolant (antifreeze).

Many repair and service procedures generate what are known as hazardous wastes. Dirty solvents and cleaners are good examples of hazardous wastes. Something is classified as a hazardous waste if it is on the Environmental Protection Agency (EPA) list of known harmful materials or has one or more of the following characteristics.

- *Ignitability.* A liquid with a flash point below 140°F or a solid that can spontaneously ignite.
- *Corrosivity.* A substance dissolves metals and other materials or burns the skin.
- *Reactivity.* Any material that reacts violently with water or other materials or releases cyanide gas, hydrogen sulfide gas, or similar gases when exposed to low-pH acid solutions. This includes materials that generate toxic mists, fumes, vapors, or flammable gases.
- *Toxicity.* Materials that leach one or more of eight heavy metals in concentrations greater than 100 times primary drinking water standard concentrations.

Complete EPA lists of hazardous wastes can be found in the Code of Federal Regulations. It should be noted that no material is considered hazardous waste until the shop is finished using it and ready to dispose of it.

The following list describes the recommended procedure for dealing with some of the common hazardous wastes. Always follow these and any other mandated procedures.

Refrigerants Recover or recycle refrigerants (or do both) during the service and disposal of motor vehicle air conditioner and refrigeration equipment. It is not allowable to knowingly vent refrigerants to the atmosphere. Recovering or recycling during servicing must be performed by an EPA-certified technician using certified equipment and following specified procedures.

Containers Cap, label, cover, and properly store above-ground and outdoors any liquid containers and small tanks within a diked area and on a paved impermeable surface to prevent spills from running into surface or ground water.

Oil Recycle oil. Set up equipment, such as a drip table or screen table with a used oil collection bucket, to collect oils dripping off parts. Place drip pans underneath vehicles that are leaking fluids onto the storage area. Do not mix other wastes with used oil, except as allowed by your recycler. Used oil generated by a shop (or oil received from household "do-it-yourself" generators) may be burned on site in a commercial space heater. Used oil also may be burned for energy recovery. Contact state and local authorities to determine requirements and to obtain necessary permits.

Oil filters Drain for at least 24 hours, crush, and recycle used oil filters.

Batteries Recycle batteries by sending them to a reclaimer or back to the distributor. Keeping shipping receipts can demonstrate that you have recycled. Store batteries in a watertight, acid-resistant container. Inspect batteries for cracks and leaks when they come in. Treat a dropped battery as if it were cracked. Acid residue is hazardous because it is corrosive and may contain lead and other toxics. Neutralize spilled acid by using baking soda or lime, and dispose of it as hazardous material.

Metal residue from machining Collect metal filings when machining metal parts. Keep separate and recycle if possible. Prevent metal filings from falling into a storm sewer drain.

Solvents Replace hazardous chemicals with less toxic alternatives that have equal performance. For example, substitute water-based cleaning solvents for petroleum-based solvent degreasers. To reduce the amount of solvent used when cleaning parts, use a two-stage process (dirty solvent followed by fresh solvent). Hire a hazardous waste management service to clean and recycle solvents. (Some spent solvents must be disposed of as hazardous waste unless recycled properly). Store solvents in closed containers to prevent evaporation. Evaporation of solvents contributes to ozone depletion and smog formation. In addition, the residue from evaporation must be treated as a hazardous waste. Properly label spent solvents and store in drip pans or in diked areas and only with compatible materials.

Other solids Store materials such as scrap metal, old machine parts, and worn tires under a roof or tarpaulin to protect them from the elements and to prevent potentially contaminated runoff. Consider recycling tires by retreading them.

Liquid recycling Collect and recycle coolants from radiators. Store transmission fluids, brake fluids, and solvents containing chlorinated hydrocarbons separately, and recycle or dispose of them properly.

Shop towels and rags Keep waste towels in a closed container marked "Contaminated shop towels only." To reduce costs and liabilities associated with disposal of used towels, which can be classified as hazardous wastes, investigate using a laundry service that is able to treat the wastewater generated from cleaning the towels.

Waste storage Always keep hazardous waste separate, properly labeled, and sealed in the recommended containers. The storage area should be covered and may need to be fenced and locked if vandalism could be a problem. Select a licensed hazardous waste hauler after seeking recommendations and reviewing the firm's permits and authorizations.

REFRIGERANT SAFETY PRECAUTIONS

- Always work in a well-ventilated and clean area. Refrigerant is colorless and invisible as a gas. Refrigerant is heavier than oxygen and will displace oxygen in a confined area. Avoid breathing the refrigerant vapors. Exposure to refrigerant may irritate your eyes, nose, and throat.
- Refrigerant evaporates quickly when it is exposed to the atmosphere. It will freeze anything it contacts. If liquid refrigerant gets in your eyes or on your skin, it can cause frostbite. Never rub your eyes or skin if you have come in contact with refrigerant. Immediately flush the exposed areas with cool water for 15 minutes and seek medical help.
- An A/C system's high pressure can cause severe injury to your eyes or skin if a hose bursts. Always wear eye protection when working around the A/C system and refrigerant. It is also advisable to wear protective gloves and clothing.
- Never use R-134a in combination with compressed air for leak testing. Pressurized R-134a, in the presence of oxygen, may form a combustible mixture. Never introduce compressed air into R-134a containers (empty or full ones), A/C systems, or A/C service equipment.
- Be careful when handling refrigerant containers. Never drop, strike, puncture, or burn the containers. Always use Department of Transportation (DOT)-approved refrigerant containers.
- Never expose A/C system components to high temperatures—heat causes the refrigerant's pressure to increase. Never expose refrigerant to an open flame—a poisonous gas (phosgene) results. If inhaled, phosgene gas can cause severe respiratory irritation. This gas also occurs when an open-flame leak detector is used. Phosgene fumes have an acrid (bitter) smell. If the refrigerant needs to be heated during service, the bottom of the refrigerant container should be placed in warm water (less than 125°F).
- Never overfill refrigerant containers. The filling level of the container should never exceed 60 percent of the container's gross weight rating. Always store refrigerant containers in temperatures below 125°F and keep them out of direct sunlight.
- Refrigerant comes in 30- and 50-pound cylinders. Remember that these drums are under considerable pressure and should be handled following precautions. Keep the drums in an upright position. Make sure that metal caps protect valves and safety plugs when the drums are not in use. Avoid dropping the drums. Handle them carefully. When transporting refrigerant, do not place containers in the vehicle's passenger compartment. R-134a should be stored in light blue containers. Refrigerant types should never be mixed. Their oils and desiccants are not compatible. If refrigerants are mixed, contamination will occur and may result in A/C system failure. Separate service equipment, including recovery/recycling machines and service

gauges, should be used for the different refrigerants. Always read and follow the instructions from the equipment manufacturer when servicing A/C systems.

HEATING AND AIR CONDITIONING TOOLS AND EQUIPMENT

Many different tools and pieces of testing and measuring equipment are used to service heating and air conditioning systems. NATEF has said that an air conditioning technician must know what they are and how and when to use them. The tools and equipment listed by NATEF are covered in the following discussion, along with the tools and equipment you will use while completing the job sheets. Although you will be using common hand tools, they are not part of this discussion. You should already know what they are and how to use and care for them.

NOTE: *Air conditioning systems are extremely sensitive to moisture and dirt. Therefore, clean working conditions are extremely important. The smallest particle of foreign matter in an air conditioning system contaminates the refrigerant, causing rust, ice, or damage to the compressor. For this reason, all replacement parts are sold in vacuum-sealed containers and should not be opened until they are to be installed in the system. If, for any reason, a part has been removed from its container for any length of time, the part must be completely flushed using only a recommended flush solvent to remove any dust or moisture that might have accumulated during storage. When the system has been open for any length of time, the entire system must be purged completely. A new receiver/drier must be installed because the element of the existing unit will have become saturated and unable to remove any moisture from the system once the system is recharged.*

Manifold Gauge Set

The manifold gauge set (Figure 2) is used when discharging, charging, evacuating, and diagnosing trouble in the system. With the new legislation on handling refrigerants, all gauge sets are required to have a valve device to close off the end of the hose so that the fitting not in use is automatically shut.

The low-pressure gauge is graduated into pounds of pressure from 1 to 120 (with cushion to 250) in 1-pound graduations, and, in the opposite direction, in inches of vacuum from 0 to 30. This is the gauge that should always be used in checking pressure on the low-pressure side of the system. The high-pressure gauge is graduated from 0 to 500 pounds pressure in 10-pound graduations. This gauge is used for checking pressure on the high-pressure side of the system.

The center manifold fitting is common to both the low and the high side and is for evacuating or adding refrigerant to the system. When this fitting is not being used, it should be capped. A test hose connected to the fitting directly under the low-side gauge is used to connect the low side of the test manifold to the low side of the system. A similar connection is found on the high side.

The gauge manifold is designed to control refrigerant flow. When the manifold test set is connected into the system (Figures 3 and 4), pressure is registered on both gauges at all times. During all tests, both the low- and high-side hand valves are in the closed position (turned inward until the valve is seated).

Figure 2 A manifold gauge set.

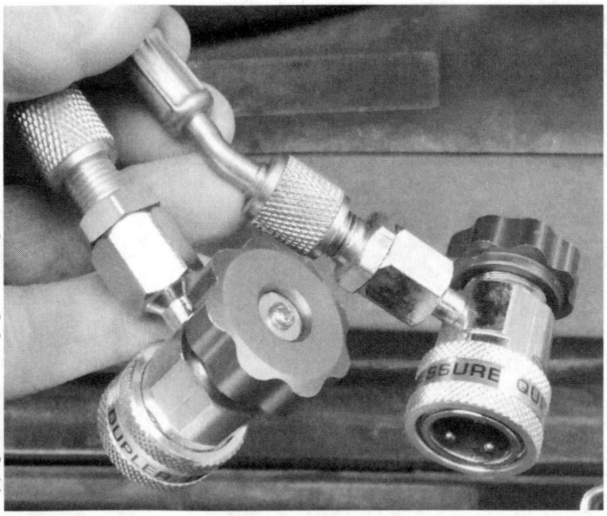

Figure 3 Low side (on left) high side (on right) gauge adapters.

Refrigerant flows around the valve stem to the respective gauges and registers the system's low-side pressure on the low-side gauge and the system's high-side pressure on the high-side gauge. The hand valves isolate the low and high sides from the central portion of the manifold. When the gauges are first connected to the gauge fittings with the refrigeration system charged, the gauge lines should always be purged. Purging is done by cracking each valve on the gauge set to allow the pressure of the refrigerant in the refrigeration system to force the air to escape through the center gauge line. Failure to purge lines can result in air or other contaminants entering the refrigeration system.

For identification purposes, R-134a service hoses must have a black stripe along their length and be clearly labeled. The low-pressure hose is blue with a black stripe. The high-pressure hose is red with a black stripe, and the center service hose is yellow with a black stripe. Service hoses for one type of refrigerant do not easily connect to the wrong system because the fittings for an R-134a system are different.

Electronic Leak Detector

An electronic leak detector is safe, effective, and can be used with all types of refrigerants. A hand-held, battery-operated electronic leak detector contains a test probe that is moved about 1 inch per second in areas of suspected leaks (Figure 5). Since refrigerant is heavier than air, the probe should be positioned below the test point. An alarm or a buzzer on the detector

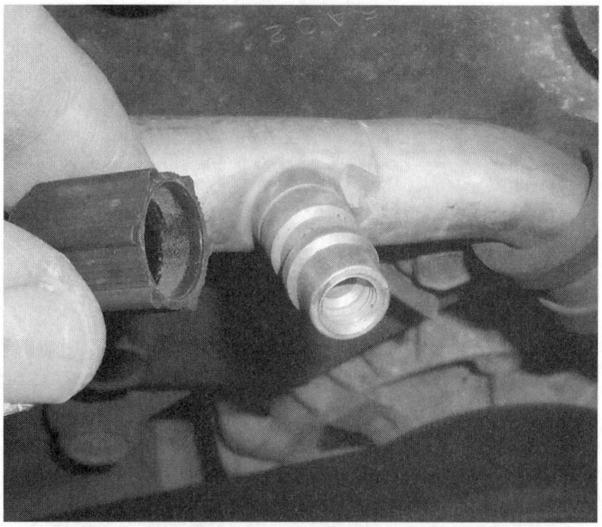

Figure 4 Low side fitting (top) High side fitting (bottom).

indicates the presence of a leak. On some models, a light flashes when refrigerant is detected.

Fluorescent Leak Tracer

To find a refrigerant leak using the fluorescent tracer system, first introduce a fluorescent dye into the air conditioning system with a special infuser included with the detector equipment. Run the air conditioner for a few minutes, giving the tracer dye fluid time to circulate and penetrate. Wearing the tracer protective goggles, scan the system with a black-light glow gun. Leaks in the system shine under the black light as a luminous yellow-green.

Dye Injectors

Dye may be added to a system by injecting it into the system or by pouring it directly into a removed component. It is important to note that all

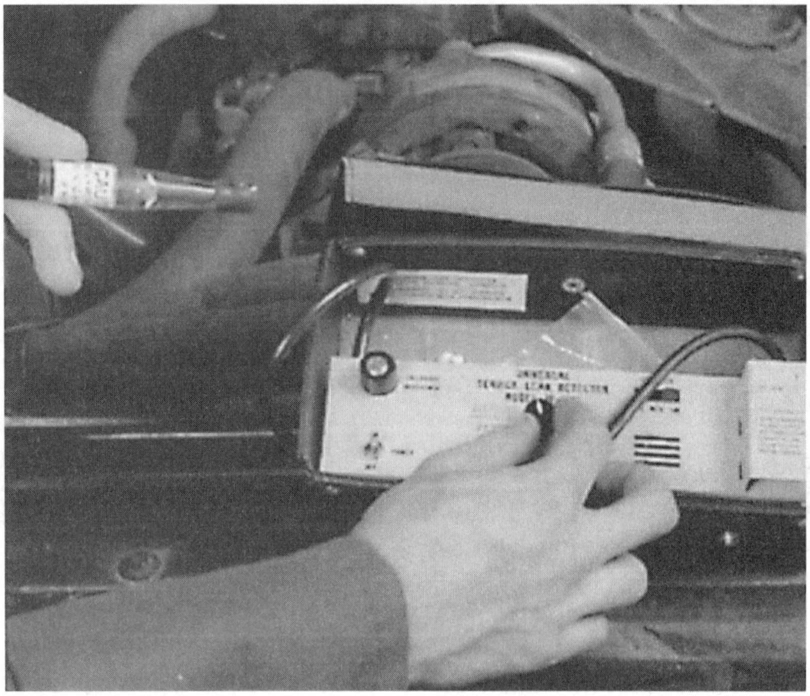

Figure 5 An electronic leak detector.

fluorescent dyes are not compatible with PAG oil. Before adding dye to the system, make sure it is the right type. Make sure the right quantity of dye is installed; never overcharge the system with dye.

There are two basic ways to inject dye into the system. One uses an A/C charging station and a manifold gauge set. This should only be used when the system is not fully charged. A small tank of dye and injector is attached to the center hose on the gauge set. The dye is added with some refrigerant to the low-side of the system.

When the system is fully charged, a special kit must be used. The kit contains a reservoir for the dye and is connected between the high-pressure and low-pressure service valves. With the A/C turned on, the high-side valve is opened and remains open until the dye has left the reservoir. This method does not add refrigerant to the system.

Leak Detector Fluid

Leaks can also be located by applying leak detector fluid around areas to be tested. If a leak is present, it will form clusters of bubbles around the source. A very small leak will cause white foam to form around the leak source within several seconds to a minute. Adequate lighting over the entire surface being tested is necessary for an accurate diagnosis.

Scan Tools

The introduction of computer-controlled systems brought with it the need for tools capable of troubleshooting electronic control systems. There are a variety of computer scan tools available today that do just that. A scan tool is a microprocessor designed to communicate with the vehicle's computer. Connected to the computer through diagnostic connectors, a scan tool can access trouble codes, run tests to check system operations, and monitor the activity of the system. Trouble codes and test results are displayed on an LED screen.

Scan tools are capable of testing many onboard computer systems, such as climate controls, transmission controls, engine computers, antilock brake computers, air bag computers, and suspension computers, depending on the year and make of the vehicle and the type of scan tester. In many cases, the technician must select the computer system to be tested with the scanner after it has been connected to the vehicle.

The scan tool is connected to specific diagnostic connectors on the vehicle. Some manufacturers have one diagnostic connector. This connects the data wire from each computer to a specific terminal in this connector. Other manufacturers have several diagnostic connectors on each vehicle, and each of these connectors may be connected to one or more computers. The scan tool must be programmed for the model year, make of vehicle, and type of engine.

With OBD-II, the diagnostic connectors are located in the same place on all vehicles. Also, any scan tool designed for OBD-II will work on all OBD-II systems, therefore the need to have designated scan tools or cartridges is eliminated. Most OBD-II scan tools have the ability to store, or "freeze" data during a road test, and then play back this data when the vehicle is returned to the shop.

There are many different scan tools available. Some are a combination of other diagnostic tools, such as a lab scope and graphing multimeter. These may have the following capabilities:

- Retrieve DTCs
- Monitor system operational data
- Reprogram the vehicle's electronic control modules
- Perform systems diagnostic tests
- Display appropriate service information, including electrical diagrams
- Display TSBs
- Troubleshooting instructions
- Easy tool updating through a personal computer (PC)

The vehicle's computer sets trouble codes when a voltage signal is entirely out of its normal range. The codes help technicians identify the cause of the problem when this is the case. If a signal is within its normal range but is still not correct, the vehicle's computer will not display a trouble code. However, a problem may still exist.

Some scan tools work directly with a PC through uncabled communication links, such as Bluetooth. Others use a Personal Digital Assistant (PDA). These are small hand-held units that allow you to read diagnostic trouble codes (DTCs), monitor the activity of sensors, and inspection/maintenance system test results to quickly determine what service the vehicle requires. Most of these scan tools also have the ability to:

- Perform system and component tests
- Report test results of monitored systems
- Exchange files between a PC and PDA
- View and print files on a PC
- Print DTC/Freeze Frame
- Generate emissions reports
- IM/Mode 6 information
- Display related TSBs
- Display full diagnostic code descriptions
- Observe live sensor data
- Update the scan tool as a manufacturer's interfaces change

DMMs

It's not necessary for a technician to own separate meters to measure volts, ohms, and amps, because a multimeter can be used instead. Top-of-the-line multimeters are multifunctional. Most test volts, ohms, and amperes in both DC and AC. Usually there are several test ranges provided for each of these functions. In addition to these basic electrical tests, multimeters also test engine rpm, duty cycle, pulse width, diode condition, frequency, and even temperature. The technician selects the desired test range by turning a control knob on the front of the meter.

Multimeters are available with either analog or digital displays. But, the most commonly used multimeter is the digital volt/ohmmeter (DVOM), which is often referred to as a digital multimeter (DMM). There are several drawbacks to using analog-type meters for testing electronic control systems. Many electronic components require very precise test results. Digital meters can measure volts, ohms, or amperes in tenths and hundredths. Another problem with analog meters is their low internal resistance (input impedance). The low input impedance allows too much current to flow through circuits and should not be used on delicate electronic devices.

Digital meters, on the other hand, have high input impedance, usually at least 10 megohms (10 million ohms). Metered voltage for resistance tests is well below 5 volts, reducing the risk of damage to sensitive components and delicate computer circuits. A high-impedance digital multimeter must be used to test the voltage of some components and systems such as an oxygen (O_2) sensor circuit. If a low-impedance analog meter is used in this type of circuit, the current flow through the meter is high enough to damage the sensor.

DMMs either have an "auto range" feature, in which the meter automatically selects the appropriate scale, or they must be set to a particular range. In either case, you should be familiar with the ranges and the different settings available on the meter you are using. To designate particular ranges and readings, meters display a prefix before the reading or range. If the meter has a setting for mAmps, this means the readings will be given in milli-amps or 1/1000th of an amp. Ohmmeter scales are expressed as a multiple of tens or use the prefix K or M. K stands for Kilo or 1000. A reading of 10K ohms equals 10,000 ohms. An M stands for Mega or 1,000,000. A reading of 10M ohms equals 10,000,000 ohms. When using a meter with an auto range, make sure you note the range being used by the meter. There is a big difference between 10 ohms and 10,000,000 ohms.

WARNING: *Many digital multimeters with auto range display the measurement with a decimal point.*

After the test range has been selected, the meter is connected to the circuit in the same way as if it were an individual meter.

When using the ohmmeter function, the DMM will show a zero or close to zero when there is good continuity. If the continuity is very poor, the meter will display an infinite reading. This reading is usually shown as a blinking "1.000," a blinking "1," or an "OL." Before taking any measurement, calibrate the meter. This is done by holding the two leads together and adjusting the meter reading to zero. Not all meters need to be calibrated; some digital meters automatically calibrate when a scale is selected. On meters that require calibration, it is recommended that the meter be zeroed after changing scales.

Multimeters may also have the ability to measure duty cycle, pulse width, and frequency. All of these represent voltage pulses caused by the turning on and off of a circuit or the increase and decrease of voltage in a circuit. Duty cycle is a measurement of the amount of time something is on compared to the time of one cycle and is measured in a percentage.

Pulse width is similar to duty cycle except that it is the exact time something is turned on and is measured in milliseconds. When measuring duty cycle, you are looking at the amount of time something is on during one cycle.

The number of cycles that occur in one second is called the frequency. The higher the frequency, the more cycles occur in a second. Frequencies are measured in Hertz. One Hertz is equal to one cycle per second.

Hybrid Tools

A hybrid vehicle is an automobile and as such is subject to many of the same problems as a conventional vehicle. Most systems in a hybrid vehicle are diagnosed in the same way as well. However, a hybrid vehicle has unique systems that require special procedures and test equipment. It is imperative to have good information before attempting to diagnose these vehicles. Also, make sure you follow all test procedures precisely as they are given.

An important diagnostic tool is a DMM. However, this is not the same DMM used on a conventional vehicle. The meter used on hybrids and other electric-powered vehicles should be classified as a category III meter. There are basically four categories for low-voltage electrical meters, each built for specific purposes and to meet certain standards. Low-voltage, in this case, means voltages less than 1000 volts. The categories define how safe a meter is when measuring certain circuits. The standards for the various categories are defined by the American National Standards Institute (ANSI), the International Electrotechnical Commission (IEC), and the Canadian Standards Association (CSA). A CAT III meter is required for testing hybrid vehicles because of the high voltages, three-phase current, and the potential for high transient voltages. Transient voltages are voltage surges or spikes that occur in AC circuits. To be safe, you should have a CAT III-1000 V meter. A meter's voltage rating reflects its ability to withstand transient voltages. Therefore, a CAT III 1000-V meter offers much more protection than a CAT III meter rated at 600 volts.

Another important tool is an insulation resistance tester. These can check for voltage leakage through the insulation of the high-voltage cables. Obviously no leakage is desired and any leakage can cause a safety hazard as well as damage to the vehicle. Minor leakage can also cause hybrid system-related driveability problems. This meter is not one commonly used by automotive technicians, but should be for any one who might service a damaged hybrid vehicle, such as doing body repair. This should also be a CAT III meter and may be capable of checking resistance and voltage of circuits like a DMM.

To measure insulation resistance, system voltage is selected at the meter and the probes placed at their test position. The meter will display the voltage it detects. Normally, resistance readings are taken with the circuit de-energized unless you are checking the effectiveness of the cable or wire insulation. In this case, the meter is measuring the insulation's effectiveness and not its resistance.

The probes for the meters should have safety ridges or finger positioners. These help prevent physical contact between your fingertips and the meter's test leads.

Lab Scopes

An oscilloscope is a visual voltmeter. An oscilloscope converts electrical signals to a visual image representing voltage changes over a specific period of time. This information is displayed in the form of a continuous voltage line called a waveform pattern or trace.

An upward movement of the voltage trace on an oscilloscope screen indicates an increase in voltage, and a downward movement of this trace represents a decrease in voltage. As the voltage trace moves across an oscilloscope screen, it represents a specific length of time.

The size and clarity of the displayed waveform is dependent on the voltage scale and the time reference selected. Most scopes are equipped with controls that allow voltage and time interval selection. It is important, when choosing the scales, to remember that a scope displays voltage over time.

Dual-trace oscilloscopes can display two different waveform patterns at the same time. This makes cause and effect analysis easier. Other scopes are capable of displaying more than two traces.

With a scope, precise measurement is possible. A scope will display any change in voltage as it occurs. This is especially important for diagnosing intermittent problems. It is also invaluable for checking the primary and secondary circuits, as well as fuel injectors and various inputs and outputs of the control system.

The screen of a lab scope is divided into small divisions of time and voltage. Time is represented by the horizontal movement of the waveform. Voltage is measured with the vertical position of the waveform. Since the scope displays voltage over time, the waveform moves from the left (the beginning of measured time) to the right (the end of measured time). The value of the divisions can be adjusted to improve the view of the voltage waveform.

Since a scope displays actual voltage, it will display any electrical noise or disturbances that accompany the voltage signal. Noise is primarily caused by radio frequency interference (RFI), which may come from the ignition system. RFI is an unwanted voltage signal that rides on a signal. This noise can cause intermittent problems with unpredictable results. The noise causes slight increases and decreases in the voltage. When a computer receives a voltage signal with noise, it will try to react to the minute changes. As a result, the computer responds to the noise rather than the voltage signal.

Graphing Multimeter

One of the latest trends in diagnostic tools is a graphing digital multimeter. These meters display readings over time, similar to a lab scope. The graph displays the minimum and maximum readings on a graph, as well as displaying the current reading. By observing the graph, a technician can identify any undesirable changes during the transition from a low reading to a high reading, or vice versa. These glitches are some of the more difficult problems to identify without a graphing meter or a lab scope.

Computer Memory Saver

Memory savers are an external power source used to maintain the memory circuits in electronic accessories and the engine, transmission, and body computers when the vehicle's battery is disconnected. The saver is plugged into the vehicle's cigar lighter outlet. It can be powered by a 9- or 12-volt battery.

Stethoscope

Some sounds can be easily heard without using a listening device, but others are impossible to hear unless amplified. A stethoscope is very helpful in locating the cause of a noise by amplifying the sound waves. It can also help you distinguish between normal and abnormal noise. The procedure for using a stethoscope is simple. Use the metal prod to trace the sound until it reaches its maximum intensity. Once the precise location has been discovered, the sound can be better evaluated. A sounding stick, which is nothing more than a long, hollow tube, works on the same principle, though a stethoscope gives much clearer results.

The best results, however, are obtained with an electronic listening device. With this tool you can tune into the noise. Doing this allows you to eliminate all other noises that might distract or mislead you.

Coolant Hydrometer

A coolant hydrometer is used to check the amount of antifreeze in the coolant. This tester contains a pickup hose, coolant reservoir, and squeeze bulb. The pickup hose is placed in the radiator coolant. When the squeeze bulb is squeezed and released, coolant is drawn into the reservoir. As coolant enters the reservoir, a pivoted float moves upward with the coolant level. A pointer on the float indicates the freezing point of the coolant on a scale located on the reservoir housing.

Refractory Testers

For many shops, the preferred way to check coolant is with a refractometer. This tester works on the principle that light bends as it passes through a liquid. A sample of the coolant is placed in the tester. As light passes through the sample of coolant, it bends and shines on a scale in the tester. A reading is taken at the point on the scale where there a separation of light and dark. Most refractory coolant testers can also check the electrolyte in a battery.

Measuring pH

Acids produced by bacteria and other contaminants can reduce the effectiveness of coolant. Some shops measure the pH of coolant to determine and deterioration of the coolant. The pH is measured by placing test strips or a digital pH tester into the coolant.

Cooling System Pressure Tester

A cooling system pressure tester contains a hand pump and a pressure gauge. A hose is connected from the hand pump to a special adapter that fits on the radiator filler neck. This tester is used to pressurize the cooling system and check for coolant leaks. Additional adapters are available to connect the tester to the radiator cap. With the tester connected to the radiator cap, the pressure-relief action of the cap may be checked.

Thermometer

A dial-type thermometer (Figure 6) is often used to measure the air temperature at the vent outlets. This procedure checks the overall performance of the system. The thermometer can also be used to check the temperature of refrigerant lines, hoses, and components while diagnosing a system. An electronic pyrometer works best for this procedure and is often used.

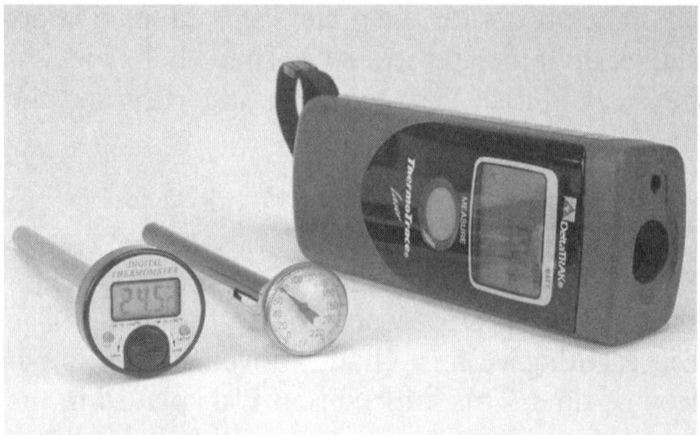

Figure 6 Three types of commonly used thermometers.

Hygrometer

A hygrometer is an instrument used to measure humidity. This instrument is basically two thermometers. One thermometer relies on a dry sensing bulb and the other is a wet bulb. The wet bulb is kept wet. Relative humidity is calculated from the ambient temperature sensed by the dry-bulb thermometer and the difference in temperatures between the wet-bulb and dry-bulb thermometers.

Feeler Gauge

A feeler gauge is a thin strip of metal or plastic of known and closely controlled thickness. Several of these strips are often assembled together as a feeler gauge set, which looks like a pocketknife. The desired thickness gauge can be pivoted away from others for convenient use. A feeler gauge set usually contains strips or leaves of 0.002- to 0.010-inch thickness (in steps of 0.001 inch) and leaves of 0.012- to 0.024-inch thickness (in steps of 0.002 inch).

A feeler gauge can be used by itself to measure clearances, such as critical compressor clutch clearances. It can also be used with a precision straightedge to measure the flatness of sealing surfaces in a compressor.

Straightedge

A straightedge is no more than a flat bar machined to be totally flat and straight; to be effective, it must be flat and straight. Any surface that should be flat can be checked with a straightedge and feeler gauge set. The straightedge is placed across and at angles on the surface. At any low points on the surface, a feeler gauge can be placed between the straightedge and the surface. The size gauge that fills in the gap is the amount of warpage or distortion.

Dial Indicator

The dial indicator is calibrated in 0.001-inch (one-thousandth-inch) increments. Metric dial indicators are also available; both types are used to measure movement. Common uses of the dial indicator include checking compressor shaft run out, shaft endplay, and compressor clutch run out.

To use a dial indicator, position the indicator rod against the object to be measured. Then, push the indicator toward the work until the indicator needle travels far enough around the gauge face to permit movement to be read in either direction. Zero the indicator needle on the gauge. Move the object in the direction required while observing the needle of the gauge. Always be sure the range of the dial indicator is sufficient to allow the amount of movement required by the measuring procedure. For example, never use a 1-inch indicator on a component that will move 2 inches.

Torque Wrench

Torque is the twisting force used to turn a fastener against the friction between the threads and between the head of the fastener and the surface of the component. The fact that practically every vehicle and engine manufacturer publishes a list of torque recommendations is ample proof of the importance of using proper amounts of torque when tightening nuts or bolts. The amount of torque applied to a fastener is measured with a torque-indicating or torque wrench.

A torque wrench is basically a ratchet or breaker bar with some means of displaying the amount of torque exerted on a bolt when pressure is applied to the handle. Torque wrenches are available with the various drive sizes. Sockets are inserted onto the drive and then placed over the bolt. As pressure is exerted on the bolt, the torque wrench indicates the amount of torque.

The common types of torque wrenches are available with inch-pound and foot-pound increments.

- A beam torque wrench is not highly accurate. It relies on a beam metal that points to the torque reading.
- A "click"-type torque wrench clicks when the desired torque is reached. The handle is twisted to set the desired torque reading.
- A dial torque wrench has a dial that indicates the torque exerted on the wrench. The wrench may have a light or buzzer that turns on when the desired torque is reached.
- A digital readout type displays the torque and is commonly used to measure turning effort, as well as for tightening bolts. Some designs of this type of torque wrench have a light or buzzer that turns on when the desired torque is reached.

Gear and Bearing Pullers

Many tools are designed for a specific purpose. An example of a special tool is a gear and bearing puller. Many gears and bearings have a slight interference fit (press fit) when they are installed on a shaft or in a housing. Something that has a press fit has an interference fit. For example, the inside diameter of a bore is 0.001 inch smaller than the outside diameter of a shaft, and when the shaft is fitted into the bore, it must be pressed in to overcome the 0.001-inch interference fit. This press fit prevents the parts from moving on each other. The removal of these gears and bearings must be done carefully to prevent damage to the gears, bearings, or shafts. Prying or hammering can break or bind the parts. A puller with the proper jaws and adapters should be used to remove gears and bearings. Using the proper puller, the force required to remove a gear or bearing can be applied with a slight and steady motion.

Bushing and Seal Pullers and Drivers

Another commonly used group of special tools are the various designs of bushing and seal drivers and pullers. Pullers are either a threaded or slide-hammer type of tool. Always make sure you use the correct tool for the job because bushings and seals are easily damaged if the wrong tool or procedure is used. Car manufacturers and specialty tool companies work together closely to design and manufacture special tools required to repair cars. Most of these special tools are listed in the appropriate service manuals.

Retaining Ring Pliers

An air conditioning technician will encounter many different styles and sizes of retaining rings and snap rings that hold things together or keep them in a fixed location. Using the correct tool to remove and install these rings is the only safe way to work with them. All technicians should have an assortment of snap ring and retaining ring pliers.

Compressor Tools

Although compressors are typically replaced when they are faulty, certain service procedures applied to them are standard practice. Most of these procedures focus on compressor clutch and shaft seal service, and they require special tools (Figure 7). Clutch plate tools are required to gain access to the shaft seal. They are also needed to reinstall the clutch plate after service.

Many different tools are required to perform these services on a compressor. Typically, to replace a shaft seal, you need an adjustable or fixed spanner wrench, clutch plate installer/remover, ceramic seal installer/remover, seal assembly installer/remover, seal seat installer/remover, shaft seal protector, snap ring pliers, O-ring remover, and O-ring installer. Some of these tools are made for a specific model compressor; others are universal fit or will have interchangeable parts to allow them to work on a variety of compressors.

Can Tap

A can tap is used to connect a 1-"pound" can of refrigerant to the manifold gauge set. These cans are available in either screw top or flat top. The can tap either pierces through the top or makes a screw-type connection with the top. It should be noted that "1-pound" cans actually contain only 12 or 14 ounces. Many states have restrictions or strict prohibitions on the sale and use of small 1-pound cans of refrigerant.

Hose and Fitting Tools

An A/C system is a closed system, meaning outside air should never enter the system, and the refrigerant in the system should never exit to the outside. To maintain its integrity, special fittings and hoses are used. Often, special tools are required when servicing the system's fittings and hoses. An example of this is the spring-lock coupling tool set; without the tool, it is impossible to separate this connector without damaging it.

Charging Cylinder

The charging cylinder is designed to mete out a desired amount of refrigerant by weight. Compensation for temperature variations is accomplished by reading the pressure on the gauge of the cylinder and dialing the plastic shroud. The calibrated chart on the shroud contains corresponding pressure readings for the refrigerant being used.

Refrigerant Identifier

A refrigerant identifier (Figure 8), obviously, is used to identify the type of refrigerant that is present in a system. This needs to be done before any service work begins. The tester is used to identify the purity and quality of the refrigerant sample taken from the system.

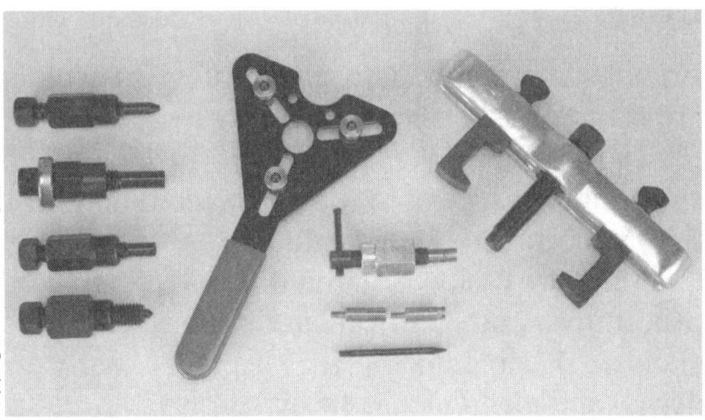

Figure 7 A sample of common compressor tools.

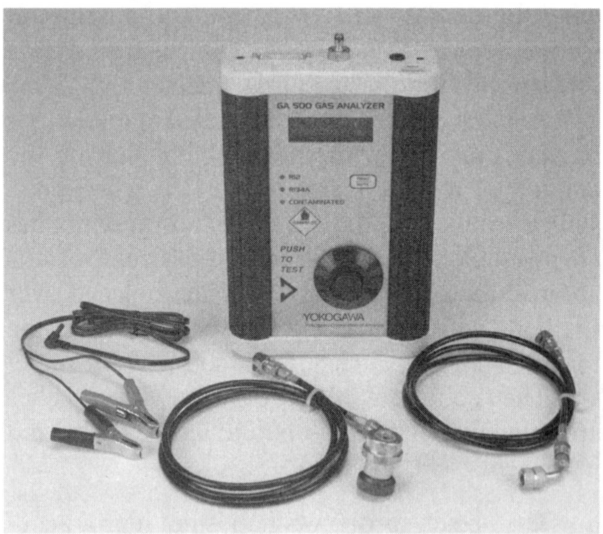

Figure 8 A refrigerant identifier.

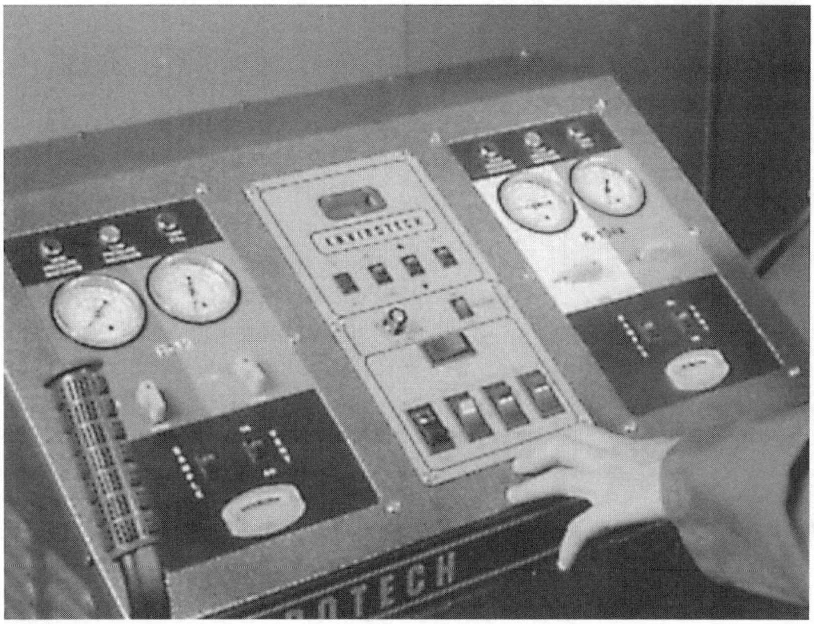

Figure 9 A refrigerant recover/recycling machine.

Refrigerant Recovery/ Recycling System

Each shop that does A/C work must have a recovery, recycle, and recharge machine (Figure 9) for each type of refrigerant.

There are currently two types of refrigerant recovery machines, the single-pass and the multipass. Both have the ability to draw the refrigerant from the vehicle, filter and separate the oil from it, remove moisture and air from it, and store the refrigerant until it is reused.

In a single-pass system, the refrigerant goes through each stage before being stored. In multipass systems, the refrigerant may go through all stages or some of the stages before being stored. Either system is acceptable if it has the UL-approved label.

All recycled refrigerant must be safely stored in DOT CFR Title 49-approved or UL-approved containers. Containers specifically made for R-134a should be so marked. Before any container of recycled refrigerant can be used, it must be checked for noncondensable gases.

Vacuum Pump

It is critical that all moisture be removed from a system before it is recharged. Moisture is removed by connecting a vacuum pump to the system. The vacuum pump may be a separate item or may be incorporated into other air conditioning equipment, such as the recovery/recycling machine.

The main purpose of the vacuum pump is to remove the contaminating air and moisture from the system. The vacuum pump reduces system pressure in order to vaporize the moisture and then exhausts the vapor along with all remaining air. The pump's ability to clean the system is directly related to its ability to reduce pressure (create a vacuum) low enough to boil off all the contaminating moisture.

Thermistor Vacuum Gauge

An electronic thermistor vacuum gauge is designed to work with the vacuum pump to measure the last, most critical inch of mercury during evacuation. It constantly monitors and visually indicates the vacuum level so that you know when the system has a full vacuum and will be free of moisture.

After the system is evacuated, it can be recharged. If the system will not pull down to a good vacuum, there is probably a leak somewhere in the system.

Electronic Scale

An electronic scale (Figure 10) is used to ensure that an accurate amount of refrigerant has been installed in a system. Since refrigerant is installed by weight, precise measurements of the refrigerant container before and after charging the system indicates how much refrigerant has been added to the system.

Figure 10 An electronic scale.

Coolant Recovery and Recycle System

A coolant recovery and recycle machine typically can drain, recycle, fill, flush, and pressure test a cooling system. Typically, additives are mixed into the used coolant during recycling. These additives either bind to contaminants in the coolant so they can be easily removed, or they restore some of the chemical properties in the coolant.

Service Information

Perhaps the most important tools you will use is service information. There is no way a technician can remember all the procedures and specifications needed to correctly repair all vehicles. Thus, a good technician relies on service information. Good information plus knowledge allows a technician to fix a problem with the least amount of frustration and at the lowest cost to the customer.

To obtain the correct system specifications and other information, you must first identify the exact system you are working on. The best source for vehicle identification is the vehicle identification number (VIN). The code can be interpreted through information given in the service information. In addition, the information may also help you recognize the system through identification of key components or other numbers or markings.

The primary source of repair and specification information for any car, van, or truck is the manufacturer. The manufacturer makes service information available to technicians through the Internet via subscription. In the case of air conditioning systems, there is a section for each engine that may be in the vehicle. Often the heating system is included with information about the engine's cooling system. Manufacturers' manuals cover all repairs, adjustments, specifications, detailed diagnostic procedures, and the special tools required.

Since many technical changes occur on specific vehicles each year, manufacturers' service information must be constantly updated. Updates are published as service bulletins (often referred to as Technical Service Bulletins, or TSBs) that show the changes in specifications and repair procedures during the model year. Service information is updated constantly through the Internet. Service information is available from independent companies, as well as from the manufacturers. However, they pay for and get most of their information from the car makers. They contain component information, diagnostic steps, repair procedures, and specifications for several car makes in one book. Information is usually condensed and is more general in nature than that in the manufacturer's manuals. The condensed format allows for more coverage in less space and, therefore, is not always specific. Many of the larger parts manufacturers have excellent guides on the various parts they manufacture or supply. They also provide updated service bulletins on their products. Other sources for up-to-date technical information are trade magazines and trade associations.

The most effective and efficient way to obtain service information is through an Internet subscription. All a technician needs to do is enter vehicle information and then move to the appropriate part or system. The appropriate information will then appear on the computer's screen. Online data can be updated instantly and requires no space for physical storage. These systems are easy to use and the information is quickly accessed and displayed. The computer's keyword, mouse, and/or light pen are used to make selections from the screen's menu. Once the information is retrieved, a technician can read it off the screen or print it out and take it to the service bay. Service information from the Internet is updated immediately.

JOB SHEETS

REQUIRED SUPPLEMENTAL TASKS (RST)

JOB SHEET 1

Shop Safety Survey

Name _______________________ Station _______________ Date _______________

NATEF Correlation

This Job Sheet addresses the following **RST** tasks:

Shop and Personal Safety:

1. Identify general shop safety rules and procedures.
8. Identify the location and use of eyewash stations.
10. Comply with the required use of safety glasses, ear protection, gloves, and shoes during lab/shop activities
11. Identify and wear appropriate clothing for lab/shop activities
12. Secure hair and jewelry for lab/shop activities.

Objective

As a professional technician, safety should be one of your first concerns. This job sheet will increase your awareness of shop safety rules and safety equipment. As you survey your shop area and answer the following questions, you will learn how to evaluate the safeness of your workplace.

Materials

Copy of the shop rules from the instructor

PROCEDURE

Your instructor will review your progress throughout this worksheet and should sign off on the sheet when you complete it.

1. Have your instructor provide you with a copy of the shop rules and procedures.

   Have you read and understood the shop rules? ☐ Yes ☐ No

2. Before you begin to evaluate your work area, evaluate yourself. Are you dressed to work safely? ☐ Yes ☐ No

   If no, what is wrong? ____

3. Are your safety glasses OSHA approved? ☐ Yes ☐ No

   Do they have side protection shields? ☐ Yes ☐ No

4. Look around the shop and note any area that poses a potential safety hazard.

   ____

   ____

   All true hazards should be brought to the attention of the instructor immediately.

5. What is the air line pressure in the shop? __________ psi

   What should it be? __________ psi

6. Where is the first-aid kit(s) kept in the work area?

   ____

   ____

7. Ask the instructor to show the location of, and demonstrate the use of, the eyewash station. Where is it and when should it be used?

   ____

   ____

8. What is the shop's procedure for dealing with an accident?

   ____

   ____

   ____

9. Explain how to secure hair and jewelry while working in the shop.

   ____

10. List the phone numbers that should be called in case of an emergency.

   ____

Problems Encountered

Instructor's Comments

JOB SHEET 2

Working in a Safe Shop Environment

Name ______________________ Station ______________ Date ______________

NATEF Correlation

This Job Sheet addresses the following **RST** tasks:

Shop and Personal Safety:

2. Utilize safe procedures for handling of tools and equipment.
3. Identify and use proper placement of floor jacks and jack stands.
4. Identify and use proper procedures for safe lift operation.
5. Utilize proper ventilation procedures for working within the lab/shop area.
6. Identify marked safety areas.
9. Identify the location of the posted evacuation routes.

Objective

This job sheet will help you work safely in the shop. Two of the basic tools a technician uses are lifts and the floor jack.

This job sheet also covers the technicians' environment.

Materials

Vehicle for hoist and jack stand demonstration

Service information

Describe the vehicle being worked on:

Year ________ Make ____________________ Model ____________________

VIN ____________________ Engine type and size ____________________

PROCEDURE

1. Are there safety areas marked around grinders and other machinery? ☐ Yes ☐ No
2. Are the shop emergency escape routes clearly marked? ☐ Yes ☐ No
3. Have your instructor demonstrate the exhaust gas ventilation system in the shop. Explain the importance of the ventilation system.

____

____

4. What types of Lifts are used in the shop?

____

____

5. Find the location of the correct lifting points for the vehicle supplied by the instructor. On the rear of this sheet, draw a simple figure showing where these lift points are.

6. Ask your instructor to demonstrate the proper use of the lift.

   Summarize the proper use of the lift.

   ____

   ____

   ____

   ____

7. Demonstrate the proper use of jack stands with the help of your instructor.

   Summarize the proper use of jack stands.

   ____

   ____

   ____

Problems Encountered

____

____

____

Instructor's Comments

____

____

____

JOB SHEET 3

Fire Extinguisher Care and Use

Name ______________________________ Station __________________ Date __________________

NATEF Correlation

This Job Sheet addresses the following **RST** task:

Shop and Personal Safety:

7. Identify the location and the types of fire extinguishers and other fire safety equipment; demonstrate knowledge of the procedures for using fire extinguishers and other fire safety equipment.

Objective

Upon completion of this job sheet, you will be able to demonstrate knowledge of the procedures for using fire extinguishers and other fire safety equipment, and identify the location of fire extinguishers in the shop.

NOTE: *Never fight a fire that is out of control or too large. Call the fire department immediately!*

1. Identify the location of the fire extinguishers in the shop.

____

____

2. Have the fire extinguishers been inspected recently? (Look for a dated tag.)

____

____

3. What types of fires are each of the shop's fire extinguishers rated to fight?

____

____

4. What types of fires should not be used with the shop's extinguishers?

____

____

5. One way to remember the operation of a fire extinguisher is to remember the term PASS. Describe the meaning of PASS on the following lines.

   a. P

   ____

   b. A

   ____

c. S

d. S

Problems Encountered

Instructor's Comments

JOB SHEET 4

Working Safely Around Air Bags

Name ______________________ Station ______________ Date ______________

NATEF Correlation

This Job Sheet addresses the following **RST** task:

Shop and Personal Safety:

13. Demonstrate awareness of the safety aspects of supplemental restraint systems (SRS), electronic brake control systems, and hybrid vehicle high-voltage circuits.

Objective

Upon completion of this job sheet, you should be able to work safely around and with air bag systems.

Tools and Materials

A vehicle(s) with air bag

Safety glasses, goggles

Service information appropriate to vehicle(s) used

Describe the vehicle being worked on:

Year ________ Make __________________ Model __________________

VIN __________________ Engine type and size __________________

PROCEDURE

1. Locate the information about the air bag system in the service information. How are the critical parts of the system identified in the vehicle?

____

____

2. List the main components of the air bag system and describe their location.

____

____

3. There are some very important guidelines to follow when working with and around air bag systems. Look through the service information to find the answers to the questions and fill in the blanks with the correct words.

   a. Wait at least ________ minutes after disconnecting the battery before beginning any service. The reserve ________ module is capable of storing enough energy to deploy the air bag for up to ________ minutes after battery voltage is lost.

b. Never carry an air bag module by its __________ or __________, and, when carrying it, always face the trim and air bag __________ from your body. When placing a module on a bench, always face the trim and air bag __________.

c. Deployed air bags may have a powdery residue on them. __________ is produced by the deployment reaction and is converted to __________ when it comes in contact with the moisture in the atmosphere. Although it is unlikely that harmful chemicals will still be on the bag, it is wise to wear __________ and __________ when handling a deployed air bag. Immediately wash your hands after handling a deployed air bag.

d. A live air bag must be __________ before it is disposed. A deployed air bag should be disposed of in a manner consistent with the __________ and manufacturer's procedures.

e. Never use a battery- or AC-powered __________, __________, or any other type of test equipment in the system unless the manufacturer specifically says to. Never probe with a __________ for voltage.

4. Explain how an air bag sensor should be handled before it is installed on the vehicle.

Problems Encountered

____

____

____

Instructor's Response

____

____

____

JOB SHEET 5

High-Voltage Hazards in Today's Vehicles

Name ______________________ Station ______________ Date ______________

NATEF Correlation

This Job Sheet addresses the following **RST** task:

Shop and Personal Safety:

14. Demonstrate awareness of the safety aspects of high-voltage circuits (such as high intensity discharge (HID) lamps, ignition systems, injection systems, etc.).

Objective

Upon completion of this job sheet, you will be able to describe some of the necessary precautions to take in performing work around high-voltage hazards such as HID headlamps and ignition systems.

HID Headlamp Precautions

Describe the vehicle being worked on:

Year ________ Make ____________________ Model ____________________

VIN ____________________ Engine type and size ____________________

Materials needed

Service information for HID headlamps

Describe general operating condition:

1. List three precautions a technician should observe when working with HID headlamp systems.

   ____

   ____

   ____

2. How many volts are necessary to initiate and maintain the arc inside the bulb of this HID system?

   ____

   ____

3. A technician must never probe with a test lamp between the HID ballast and the bulb. Explain why?

   ____

High-Voltage Ignition System Precautions

4. High-voltage ignition systems can cause serious injury, especially for those who have heart problems. Name at least three precautions to take when working around high-voltage ignition systems.

____

____

____

____

Problems Encountered

____

____

____

Instructor's Comments

____

____

____

JOB SHEET 6

Hybrid High-Voltage and Brake System Pressure Hazards

Name ______________________ Station ______________ Date ______________

NATEF Correlation

This Job Sheet addresses the following **RST** task:

Shop and Personal Safety:

13. Demonstrate awareness of the safety aspects of supplemental restraint systems (SRS), electronic brake control systems, and hybrid vehicle high-voltage circuits.

Objective

Upon completion of this job sheet, you will be able to describe some of the necessary precautions to take while performing work around high-voltage systems such as that used in hybrid vehicles, as well as the high-pressure systems of some braking control systems.

Tools and Materials

Appropriate service information

AUTHOR'S NOTE: *According to the vehicles' manufacturers, a technician should not work on hybrid vehicles without having the specific training, which is beyond the scope of this job sheet. This job sheet assumes that the student will be accessing information only, not actually working on live high-voltage vehicles.*

Protective Gear

Goggles or safety glasses with side shields

High-voltage gloves with properly inspected liners

Orange traffic cones to warn others in the shop of a high-voltage hazard

Describe the vehicle being worked on:

Year ________ Make ____________________ Model ____________________

VIN ____________________ Engine type and size ____________________

Describe general operating condition:

____

PROCEDURE

High Pressure Braking Systems:

NOTE: *Many vehicles have a high pressure accumulator or a high pressure pump in their braking systems. Opening one of these systems can be hazardous due to the high pressures involved.*

1. Research the vehicle you have been assigned and describe the procedure that must be followed BEFORE opening the hydraulic braking system.

____

____

____

Hybrid Vehicles:

1. Special gloves with liners are to be inspected before each use when working on a hybrid vehicle.

   A. How should the integrity of the gloves be checked?

   ____

   ____

   B. How often should the gloves be tested (at a lab) for recertification?

   ____

   C. What precautions must be taken when storing the gloves?

   ____

   ____

   ____

   D. When must the gloves be worn?

   ____

   ____

   ____

2. What color are the high-voltage cables on hybrid vehicles?

____

3. What must be done BEFORE disconnecting the main voltage supply cable?

____

____

____

4. Describe the safety basis of the "one hand rule."

____

____

____

5. Explain the procedure to disable high voltage on the vehicle you selected.

____

____

____

6. Explain the procedure to test for high voltage to ensure that the main voltage is disconnected.

____

____

Problems Encountered

____

____

____

Instructor's Comments

____

____

____

JOB SHEET 7

Material Data Safety Sheet Usage

Name ______________________________ Station __________________ Date __________________

NATEF Correlation

This Job Sheet addresses the following **RST** task:

Shop and Personal Safety:

12. Locate and demonstrate knowledge of material safety data sheets (MSDS).

Objective

On completion of this job sheet, the student will be able to locate the MSDS folder and describe the use of an MSDS sheet on the job site.

Materials

Selection of chemicals from the shop

MSDS sheets

1. Locate the MSDS folder in the shop. It should be in a prominent location. Did you have any problems finding the folder?

____

____

2. Pick a common chemical from your tool room such as brake cleaner. Locate the chemical in the MSDS folder. What chemical did you choose?

____

____

3. What is the flash point of the chemical? ___________

4. Briefly describe why the flash point is important.

____

____

5. What is the first aid if the chemical is ingested?

____

____

6. Can this chemical be absorbed through the skin?

____

____

7. What are the signs of exposure to the chemical you selected?

____

____

8. What are the primary health hazards of the chemical?

____

____

9. What is the first aid procedure for exposure to this chemical?

____

____

10. What are the recommendations for protective clothing?

____

____

Problems Encountered

____

____

____

Instructor's Comments

____

____

____

JOB SHEET 8

Measuring Tools and Equipment Use

Name ____________________________ Station ________________ Date ________________

NATEF Correlation

This Job Sheet addresses the following **RST** tasks:

Tools and Equipment:

1. Identify tools and their usage in automotive applications.
2. Identify standard and metric designations.
3. Demonstrate safe handling and use of appropriate tools.
4. Demonstrate proper cleaning, storage, and maintenance of tools and equipment.
5. Demonstrate proper use of precision measuring tools (i.e., micrometer, dial-indicator, dial caliper).

Objective

Upon completion of this job sheet, you will be able to make measurements using micrometers, dial indicators, pressure gauges, and other measuring tools. You will also be able to demonstrate the safe handling and use of appropriate tools, and demonstrate their proper use.

Tools and Materials

Items to measure selected by the instructor

Precision measuring tools: micrometer (digital or manual), dial caliper, vacuum or pressure gauge (selected by the instructor)

PROCEDURE

Have your instructor demonstrate the measuring tools that are available to you in your shop. The tools should include both standard and metric tools.

1. Describe the measuring tools you will be using.

   ____

   ____

2. Describe the items that you will be measuring.

   ____

   ____

3. Describe any special handling or safety procedures that are necessary when using the tools.

   ____

   ____

4. Describe any special cleaning or storage that these tools might require.

____

____

5. Describe the metric unit of measurement of the tools, such as millimeters, centimeters, kilopascals, etc.

____

____

6. Describe the standard unit of measure of the tools, such as inches, pounds per square inch, etc.

____

____

7. Measure the components, list them, and record your measurements in the following table.

Item measured	*Measurement taken*	*Unit of measure*

8. Clean and store the tools. Describe the process next.

Problems Encountered

____

____

____

Instructor's Comments

____

____

____

JOB SHEET 9

Preparing the Vehicle for Service and Customer

Name ______________________ Station ______________ Date ______________

NATEF Correlation

This Job Sheet addresses the following **RST** tasks:

Preparing Vehicle for Service:

1. Identify information needed and the service requested on a repair order.
2. Identify purpose and demonstrate proper use of fender covers, mats.
3. Demonstrate use of the three C's (concern, cause, and correction).
4. Review vehicle service history.
5. Complete work order to include customer information, vehicle identifying information, customer concern, related service history, cause, and correction.

Preparing Vehicle for Customer:

1. Ensure vehicle is prepared to return to customer per school/company policy (floor mats, steering wheel cover, etc.).

Objective

Upon completion of this job sheet, you will be able to prepare a service work order based on customer input, vehicle information, and service history. The student will also be able to describe the appropriate steps to take to protect the vehicle and delivering the vehicle to the customer after the repair.

Tools and Materials

An assigned vehicle or the vehicle of your choice

Service work order or computer-based shop management package

Parts and labor guide

Work Order Source: Describe the system used to complete the work order. If a paper repair order is being used, describe the source.

____

____

PROCEDURE

1. Prepare the shop management software for entering a new work order or obtain a blank paper work order. Describe the type of repair order you are going to use.

____

____

2. Enter customer information, including name, address, and phone numbers onto the work order. Task Completed ☐

3. Locate and record the vehicle's VIN. Where did you find the VIN?

____

4. Enter the necessary vehicle information, including year, make, model, engine type and size, transmission type, license number, and odometer reading. Task Completed ☐

5. Does the VIN verify that the information about the vehicle is correct?

____

6. Normally, you would interview the customer to identify his or her concerns. However, to complete this job sheet, assume the only concern is that the customer wishes to have the front brake pads replaced. Also, assume no additional work is required to do this. Add this service to the work order. Task Completed ☐

7. Prepare the vehicle for entering the service department. Add floor mats, seat covers, and steering wheel covers to the vehicle. Task Completed ☐

8. The history of service to the vehicle can often help diagnose problems, as well as indicate possible premature part failure. Gathering this information from the customer can provide some of the data needed. For this job sheet, assume the vehicle has not had a similar problem and was not recently involved in a collision. Service history is further obtained by searching files for previous service. Often this search is done by customer name, VIN, and license number. Check the files for any related service work. Task Completed ☐

9. Search for technical service bulletins on this vehicle that may relate to the customer's concern. Did you find any? __________ If so, record the reference numbers here.

____

____

10. Based on the customer's concern, service history, TSBs, and your knowledge, what is the likely cause of this concern?

____

____

11. Add this information to the work order. Task Completed ☐

12. Prepare to make a repair cost estimate for the customer. Identify all parts that may need to be replaced to correct the concern. List these here.

____

____

13. Describe the task(s) that will be necessary to replace the part.

____

____

14. Using the parts and labor guide, locate the cost of the parts that will be replaced and enter the cost of each item onto the work order at the appropriate place for creating an estimate. If the valve or cam cover is leaking, what part will need to be replaced?

____

15. Now, locate the flat rate time for work required to correct the concern. List each task with its flat rate time.

____

____

16. Multiply the time for each task by the shop's hourly rate and add the cost of each item to the work order at the appropriate place for creating an estimate. Ask your instructor which shop labor rate to use and record it here.

____

17. Many shops have a standard amount they charge each customer for shop supplies and waste disposal. For this job sheet, use an amount of ten dollars for shop supplies. Task Completed ☐

18. Add the total costs and insert the sum as the subtotal of the estimate. Task Completed ☐

19. Taxes must be included in the estimate. What is the sales tax rate and does it apply to both parts and labor, or just one of these?

____

____

20. Enter the appropriate amount of taxes to the estimate, then add this to the subtotal. The end result is the estimate to give the customer. Task Completed ☐

21. By law, how accurate must your estimate be?

____

____

22. Generally speaking, the work order is complete and is ready for the customer's signature. However, some businesses require additional information; make sure you add that information to the work order. On the work order, there is a legal statement that defines what the customer is agreeing to. Briefly describe the contents of that statement.

____

23. Now that the vehicle is completed and it is ready to be returned to the customer, what are the appropriate steps to take to deliver the vehicle to the customer? What would you do to make the delivery special? What should not happen when the vehicle is delivered to the customer?

____

____

Problems Encountered

____

____

____

Instructor's Comments

____

____

____

JOB SHEETS

HEATING AND AIR CONDITIONING JOB SHEET 10

Identifying Problems and Concerns

Name ______________________ Station ______________ Date ______________

NATEF Correlation

This Job Sheet addresses the following **AST/MAST** task:

A.1. Identify and interpret heating and air conditioning concern; determine necessary action.

Objective

Upon completion of this job sheet, you will be able to define heating and air conditioning system problems or concerns prior to diagnosing or testing the systems.

Protective Clothing

Goggles or safety glasses with side shields

Describe the vehicle being worked on:

Year ________ Make ____________________ Model ____________________

VIN ____________________ Engine type and size ____________________

PROCEDURE

1. Start the engine and describe how the engine seems to be running.

____

____

2. Move the temperature control on the dash to its full heat position. Describe the amount of heat coming from the heater in each of the fan positions.

____

____

3. Were there any unusual noises or smells when the heater fan was running? If so, describe them and when they occurred.

____

____

4. Turn the temperature control to the defrost mode. Describe the amount of heat coming from the defroster in each of the fan positions.

____

____

5. Were there any unusual noises or smells when the heater fan was running? If so, describe them and when they occurred.

____

____

6. Turn the temperature control to cool or the air conditioning position. Describe what the engine did when the A/C was turned on.

____

____

7. Describe how the air feels in each of the various fan positions.

____

____

8. Were there any unusual noises or smells when the A/C fan was running? If so, describe them and when they occurred.

____

____

9. Summarize your findings in detail. Include both improper and proper system operation.

____

____

____

____

10. Based on the preceding information, what are your suspicions and conclusions?

____

____

____

Problems Encountered

Instructor's Comments

HEATING AND AIR CONDITIONING JOB SHEET 11

Gathering Vehicle Information

Name ______________________ Station ______________ Date ______________

NATEF Correlation

This Job Sheet addresses the following **MLR** task:

A.1. Research applicable vehicle and service information, fluid type, vehicle service history, service precautions, and technical service bulletins.

This Job Sheet addresses the following **AST/MAST** task:

A.2. Research applicable vehicle and service information, fluid type, vehicle service history, service precautions, and technical service bulletins.

Objective

Upon completion of this job sheet, you will be able to gather service information about a vehicle and its heating and air conditioning system.

Tools and Materials

Appropriate service information

Computer

Protective Clothing

Goggles or safety glasses with side shields

Describe the vehicle being worked on:

Year ________ Make ____________________ Model ____________________

VIN ____________________ Engine type and size ____________________

PROCEDURE

1. Using the service information or other information source, describe what each letter and number in the VIN for this vehicle represents.

____

____

2. While looking in the engine compartment did you find a label regarding the vehicle's air conditioning system? Describe where you found it.

____

____

3. Summarize the information contained on this label.

____

____

4. Using the service information, locate the section about the vehicle's air conditioning system. List the major components of the system and describe how the system's pressure is controlled.

____

____

____

____

____

From the service information, locate and record all service precautions regarding the air conditioning system noted by the manufacturer.

____

____

____

____

5. Using the information available, locate and record the vehicle's service history (if possible).

____

____

____

____

6. Using the information sources available, summarize all Technical Service Bulletins for this vehicle that relate to the heating and air conditioning system.

____

____

____

____

Problems Encountered

____

____

____

Instructor's Comments

____

____

____

HEATING AND AIR CONDITIONING JOB SHEET 12

Conducting a Performance Test on an Air Conditioning System

Name ______________________________ Station __________________ Date __________________

NATEF Correlation

This Job Sheet addresses the following **AST/MAST** task:

A.3. Performance test A/C system; identify problems.

Objective

Upon completion of this job sheet, you will be able to conduct an A/C system performance test and use theory to help diagnose basic problems within the system.

Tools and Materials

Service information

Thermometer

Protective Clothing

Goggles or safety glasses with side shields

Describe the vehicle being worked on:

Year __________ Make _________________________ Model _________________________

VIN _________________________ Engine type and size _________________________

PROCEDURE

1. Describe the type of air conditioning system that is on this vehicle.

____

____

____

____

2. Carefully inspect the condenser for damage and check for anything that may block air flow. Describe your findings.

____

____

3. Inspect the drive belts and hoses for the A/C system and describe your findings.

____

____

4. Check the mounting of the compressor and describe your findings.

____

____

5. Start the engine and operate the heating and air conditioning control in each of its positions. List any switch position that the system did not respond to correctly.

____

____

6. Keeping your hands away from the cooling fan, measure the temperature of the air in front of the condenser. What was your measurement?

____

7. Place the thermometer in the center duct in the dash and set the A/C control to its maximum position and the fan to its highest speed. Raise the speed of the engine to about 1500 rpm and note the reading on the thermometer. What was it?

____

8. What reading would be satisfactory? Why? Explain why outside ambient temperature will affect the temperature of the air coming out of the duct.

____

____

____

____

9. Listen to the system and record any unusual sounds. Be sure to listen to the system while the compressor clutch cycles on and off.

____

____

10. Carefully place your hand on the evaporator near the refrigerant inlet. Describe what you feel.

____

____

11. Carefully place your hand on the evaporator near the refrigerant outlet. Describe what you feel.

____

____

12. Was there a change in temperature between the inlet and the outlet?

____

13. Based on the preceding inspection and performance test, what are your conclusions and recommendations?

____

____

____

Problems Encountered

____

____

____

Instructor's Comments

____

____

____

HEATING AND AIR CONDITIONING JOB SHEET 13

Conducting a Visual Inspection of an Air Conditioning System

Name ____________________ Station ____________________ Date ____________________

NATEF Correlation

This Job Sheet addresses the following **AST/MAST** task:

A.4. Identify abnormal operating noises in the A/C system; determine necessary action.

Objective

Upon completion of this job sheet, you will be able to conduct a thorough visual inspection of an air conditioning system before beginning detailed testing of the system.

Tools and Materials

Service information

Protective Clothing

Goggles or safety glasses with side shields

Describe the vehicle being worked on:

Year __________ Make ____________________ Model ____________________

VIN ____________________ Engine type and size ____________________

Describe the type of air conditioning used on the vehicle:

____

PROCEDURE

1. Check the condition of the compressor's drive belt. Describe its condition.

   ____

   ____

   ____

2. Check the refrigerant hose and fittings from the:

   a. Compressor to the condenser

   b. Receiver/drier to the accumulator

   c. Condenser to the evaporator

   d. Evaporator to the compressor

Describe your findings.

____

____

____

3. Check all electrical connections to the system. Describe your findings.

____

____

____

4. Start the engine and turn on the air conditioning system. After the system has been on for a few minutes, check the condenser by feeling up and down the face or along the return bends for a temperature change. Describe your findings.

____

____

____

There should be a gradual change from hot to warm as you go from the top to the bottom. Any abrupt change indicates a restriction, and the condenser has to be flushed or replaced.

5. If the system has a receiver/drier, check it. The inlet and outlet lines should be the same temperature. Any temperature difference or frost on the lines or receiver tank is a sign of a restriction. Describe your findings.

____

____

____

6. Feel the liquid line from the receiver/drier to the expansion valve. The line should be warm for its entire length. Describe your findings.

____

____

____

7. The expansion valve should be free of frost, and there should be a sharp temperature difference between its inlet and outlet. Describe your findings.

____

____

____

8. The suction line to the compressor should be cool to the touch from the evaporator to the compressor. If it is covered with thick frost, that might indicate that the expansion valve is flooding the evaporator. Describe your findings.

____

____

____

9. Check the entire system for the presence of frost. Typically, the formation of frost on the outside of a line or component means there is a restriction to the flow of refrigerant. Describe your findings.

____

____

____

10. On vehicles equipped with the orifice tube system, feel the liquid line from the condenser outlet to the evaporator inlet. A restriction is indicated by any temperature change in the liquid line before the crimp dimples the orifice tube in the evaporator inlet. Describe your findings.

____

____

____

11. Based on the visual inspection, what are your conclusions about this air conditioning system?

____

____

____

Problems Encountered

____

____

____

Instructor's Comments

____

____

____

HEATING AND AIR CONDITIONING JOB SHEET 14

Using a Pressure Gauge Set

Name ______________________ Station ______________ Date ______________

NATEF Correlation

This Job Sheet addresses the following **AST/MAST** task:

A.5. Identify refrigerant type; select and connect proper gauge set; record temperature and pressure readings.

Objective

Upon completion of this job sheet, you will be able to use service information and the labeling in the engine compartment to identify the type of air conditioning used in a vehicle and check the system's overall performance. You will also be able to correctly connect a pressure gauge set to an air conditioning system and record pressures.

Tools and Materials

Service information

Manifold gauge set

Protective Clothing

Goggles or safety glasses with side shields

Describe the vehicle being worked on:

Year ________ Make ____________________ Model ____________________

VIN ____________________ Engine type and size ____________________

PROCEDURE

1. Look at the instrument panel controls for the air-conditioning system. Describe the controls.

____

____

2. Look at all of the labeling in the engine compartment and locate any label that addresses the air conditioning system. Summarize the contents of the labeling.

____

____

3. What refrigerant is this system designed to use? ______________

What type of refrigerant oil must be used? ______________

4. Locate the information about this vehicle's air conditioning system in the service information. Record the exact sources where the information was found.

___

___

___

5. Read through the material and determine what controls the flow of refrigerant through the evaporator. Summarize your findings here.

___

___

___

6. Locate the low-side service valve and describe where it is.

___

___

7. Locate the high-side service valve and describe its location.

___

___

8. Look carefully at the fittings used at the compressor. What types of fittings are they?

___

___

9. Gather the appropriate manifold and gauge set. What color are the hoses on the gauge set and what do the colors indicate?

___

___

10. Make sure the valves on the gauge set are fully closed. Task Completed ☐

11. Remove the protective cap from the low-side service valve. Task Completed ☐

12. Connect the low-side service hose to the low-side service valve. Task Completed ☐

13. Remove the protective cap from the high-side service valve. Task Completed ☐

14. Connect the low-side service hose to the high-side service valve. Task Completed ☐

15. Start the engine and set it to run at about 1500 rpm. How did you set this speed?

___

___

16. Place a fan in front of the radiator to provide adequate airflow. Task Completed ☐

17. Turn on the air conditioning system with all the controls set to the maximum cooling position. Task Completed ☐

18. Observe the pressures on both gauges. What were they?

____

____

19. Compare your readings to specifications and describe what they indicate.

____

____

20. Return the engine to a normal idle speed. Task Completed ☐

21. Turn off the A/C and the engine. Task Completed ☐

22. Close the service hose valves, if so equipped. Task Completed ☐

23. Disconnect the service hoses. Task Completed ☐

24. Reinstall the protective caps on the service valves. Task Completed ☐

25. If the system used a different refrigerant, how would you expect the outlet temperature and system pressures to be different?

____

____

____

____

____

____

Problems Encountered

____

____

____

Instructor's Comments

____

____

____

HEATING AND AIR CONDITIONING JOB SHEET 15

Leak Testing an A/C System

Name ______________________ Station ______________ Date ______________

NATEF Correlation

This Job Sheet addresses the following **AST/MAST** task:

A.6. Leak test A/C system; determine necessary action.

Objective

Upon completion of this job sheet, you will be able to leak test an air conditioning system.

Tools and Materials

Leak detection dye

Ultraviolet black lamp

Electronic leak detector

Protective Clothing

Goggles or safety glasses with side shields

Describe the vehicle being worked on:

Year ________ Make ____________________ Model ____________________

VIN ____________________ Engine type and size ____________________

PROCEDURE

1. Inspect the system and look for oily residue on hoses, connections, and sealing points. Summarize the results of your check.

____

____

2. All suspected leak locations should be verified by using leak detection equipment. The following procedures are based on the two most common ways of leak detection. Follow the one that applies to the equipment you will be using. What equipment will you use to locate leaks?

____

____

Dye Leak Detection

1. Attach the manifold gauge set to the system's fittings. Task Completed ☐
2. Connect the center hose of the gauge set to one end of the dye injector and the other end to a can of refrigerant. Open the injector and add the refrigerant to the system. If the dye is pre-mixed with the refrigerant, add the mixture as a liquid to the system. Task Completed ☐
3. Operate the A/C system for at least 15 minutes, and then turn off the A/C and the engine. Task Completed ☐
4. Using the ultraviolet black lamp, inspect all fittings and joints. Summarize the results of your check.

   ____

   ____

5. If there was no evidence of leakage, wait 24 hours and check again. Task Completed ☐

Electronic Leak Detection

1. Refer to the manufacturer's instructions and calibrate the leak detector. Task Completed ☐
2. Slowly move the probe of the detector along the refrigerant lines and A/C components. Keep the probe about one-quarter inch from the surface of the lines and components. The beeping of the tester will increase as it gets closer to the leak. Summarize the results of your check.

   ____

   ____

3. If you detect a leak, verify the leak by using shop air to clear the immediate area around the suspected leak. Then, test for leaks again. Summarize the results of your check.

   ____

   ____

Problems Encountered

____

____

____

Instructor's Comments

____

____

____

HEATING AND AIR CONDITIONING JOB SHEET 16

Working with Refrigerant Oils

Name ______________________ Station ______________ Date ______________

NATEF Correlation

This Job Sheet addresses the following **AST/MAST** tasks:

A.7. Inspect the condition of refrigerant oil removed from the system; determine necessary action.

A.8. Determine recommended oil and oil capacity for system application.

Objective

Upon completion of this job sheet, you will be able to inspect the condition of discharged oil and select the correct type of oil for a system. You will also be able to measure and add oil to an A/C system.

Tools and Materials

Graduated container

Service information

Protective Clothing

Goggles or safety glasses with side shields

Describe the vehicle being worked on:

Year __________ Make ______________________ Model ______________________

VIN ______________________ Engine type and size ______________________

PROCEDURE

1. The compressor oil level is checked only where there is evidence of a major loss of system oil that could be caused by a broken refrigerant hose, severe hose fitting leak, badly leaking compressor seal, or collision damage to the system's components. Is this the situation? Note: many compressors do not have a provision to check the compressor oil.

   ____

   ____

2. When replacing refrigerant oil, it is important to use the specific type and quantity of oil recommended by the compressor manufacturer. If there is a surplus of oil in the system, too much oil circulates with the refrigerant, causing the cooling capacity of the system to be reduced. Too little oil results in poor lubrication of the compressor. When there has been excessive leakage or it is necessary to replace a component of the refrigeration system, certain procedures must be followed to ensure that the total oil charge in the system is correct after leak repair or the new part is on the car. Is this the situation?

   ____

   ____

3. To maintain the original total oil charge, it is necessary to compensate for this by adding oil to the new replacement part. Because of the differences in compressor designs, be sure to follow the manufacturer's instructions when adding refrigerant oil to their unit. Refer to the service information and give a summary of the amount of oil that should be added for each component that may be replaced.

____

____

____

____

4. Name the type of oil that should be used in this system. R-134a systems use synthetic polyalkaline glycol (PAG) oils. Using a mineral oil with R-134a will result in A/C compressor failure because of poor lubrication. Use only the oil specified for the system.

____

5. If a component is being replaced and you are unsure how much should be added to the system with the new component, completely drain the oil from the old component and measure it in a graduated container. Add that amount to the component when installing it. When do you add this oil?

____

____

Problems Encountered

____

____

____

Instructor's Comments

____

____

____

HEATING AND AIR CONDITIONING JOB SHEET 17

Using a Scan Tool on HVAC

Name ______________________ Station ______________ Date ______________

NATEF Correlation

This Job Sheet addresses the following **AST/MAST** task:

A.9. Using a scan tool, observe and record related HVAC data and trouble codes.

Objective

Upon completion of this job sheet, you will be able to connect a scan tool and retrieve and diagnose the HVAC system.

Tools and Materials

Service information

Scan tool

Protective Clothing

Goggles or safety glasses with side shields

Describe the vehicle being worked on:

Year ________ Make ____________________ Model ____________________

VIN ____________________ Engine type and size ____________________

PROCEDURE

Using a Scan Tool

1. Using the service information, briefly describe the modules in the network that are related to the HVAC system.

____

____

____

____

2. Connect the scan tool. Task Completed ☐
3. Turn the ignition on. Task Completed ☐
4. Program the scan tool for that vehicle. Task Completed ☐

5. Retrieve and record all DTCs. What DTCs were observed and what do they indicate?

____

____

6. To observe the activity of the inputs and outputs, select DATA. Were any of the sensors or actuators operating out of their normal range? Which ones?

____

____

7. Turn the ignition off. Task Completed ☐

8. Disconnect the scan tool. Task Completed ☐

Using the Scan Tool for Diagnosis

1. What steps will you take to diagnose the DTCs that were stored in the HVAC system?

____

____

____

____

____

____

____

____

2. Record all DTCs that were displayed and describe what is indicated by each.

____

____

3. If there are no DTCs detected and the system is still not operating properly, check the sensor input to the HVAC control unit. Task Completed ☐

4. Do any of the sensor readings seem to be incorrect?

____

____

5. After completing all repairs to the system, check to make sure the system is operating properly. Is it operating normally?

____

____

Problems Encountered

Instructor's Comments

HEATING AND AIR CONDITIONING JOB SHEET 18

Servicing Drive Belts

Name ______________________ Station ______________ Date ______________

NATEF Correlation

This Job Sheet addresses the following **MLR/AST/MAST** task:

B.1. Inspect and replace A/C compressor drive belts, pulleys, and tensioners; determine necessary action.

Objective

Upon completion of this job sheet, you will be able to inspect, replace, and adjust the drive belt for an A/C compressor and its associated parts.

Tools and Materials

Service information

Straightedge

Pry bar

Protective Clothing

Goggles or safety glasses with side shields

Describe the vehicle being worked on:

Year ________ Make ____________________ Model ____________________

VIN ____________________ Engine type and size ____________________

PROCEDURE

Serpentine Belts

1. Check the condition of the drive belt. Carefully look to see if it is cracked or glazed. What is the condition of the belt?

   ____

   ____

2. If the belt needs to be replaced, disconnect the electric cooling fan at the radiator, if the vehicle has one. Task Completed ☐

3. Always use the exact replacement size of belt. The size of a new belt is typically given, along with the part number, on the belt container. What is the part number?

   ____

4. Locate a belt routing diagram in the service information or on an underhood decal. Where did you find this diagram?

____

5. Compare the diagram with the routing of the old belt. If the actual routing is different from the diagram, draw the existing routing here.

Task Completed ☐

6. Loosen the tension on the old belt and remove the belt. How did you relieve the tension?

____

7. Inspect the grooves of the drive pulleys for rust, oil, wear, and other damage. What did you find and what service do you recommend?

____

____

8. Check the alignment of the pulleys using the straightedge. What service do you recommend?

____

____

9. Make sure the belt tensioner or idler pulley is working properly. This pulley may be a spring-loaded tensioner or an adjustable pulley. Describe its condition.

____

____

10. After all the parts are determined to be in good order, install the new belt. Make sure to wrap it according to instructions. Also, make sure the ribs of the belt are seated in the matched grooves on the pulleys.

11. Once the belt is fully routed, put tension on the belt and adjust it to specifications. What are the specifications?

____

____

Problems Encountered

___

___

___

Instructor's Comments

___

___

___

HEATING AND AIR CONDITIONING JOB SHEET 19

Inspect, Test, and Service a Compressor Clutch

Name ______________________ Station ______________ Date ______________

NATEF Correlation

This Job Sheet addresses the following **AST/MAST** task:

B.2. Inspect, test, and/or replace A/C compressor clutch components and/or assembly; check compressor clutch air gap and adjust as needed.

Objective

Upon completion of this job sheet, you will be able to inspect, test, and service a compressor clutch.

Tools and Materials

Feeler gauge set
DMM
Necessary pullers and installers
Service information

Protective Clothing

Goggles or safety glasses with side shields

Describe the vehicle being worked on:

Year __________ Make ______________________ Model ______________________

VIN ______________________ Engine type and size ______________________

Describe the type of air conditioning found on the vehicle:

____

PROCEDURE

1. Visually inspect the compressor clutch for damage and signs or oil leakage. Describe your findings.

____

____

2. Check all electrical connections in the entire A/C system. Describe your findings.

____

____

3. Check the electrical connections to the clutch. Describe your findings.

____

____

4. Using a test light or DMM with the AC system on, check the electrical plug at the clutch for power. Describe your findings.

____

____

5. If there was no power, locate and repair the problem. Retest. Describe your findings.

____

____

6. When power is present, check the ground circuit with a DMM. Describe your findings.

____

____

7. If it is determined that both power and ground are present at the compressor clutch, the clutch is defective and must be replaced. Describe your findings.

____

____

8. On some vehicles, the clutch can be serviced on the vehicle. On others, the compressor assembly must be removed to service the clutch. Refer to the service information on this vehicle to determine what must be done. Describe the procedure.

____

____

____

____

9. When replacing any part of the clutch assembly or when there is undesirable noise, check the air gap at the clutch. Locate the compressor clutch air gap specifications. The specified gap is: ____________________

10. Measure the current gap with a feeler gauge.

    The measured gap is: __________

    If the gap is outside of specifications, what should you do?

11. Test the clutch operation after the repairs are completed. Did it operate correctly and without noise?

____

____

Problems Encountered

Instructor's Comments

HEATING AND AIR CONDITIONING JOB SHEET 20

R&R a Compressor and Its Mountings

Name ______________________ Station ______________ Date ______________

NATEF Correlation

This Job Sheet addresses the following **AST/MAST** task:

B.3. Remove, inspect, and reinstall A/C compressor and mountings; determine required oil quantity.

Objective

Upon completion of this job sheet, you will be able to remove and replace an A/C compressor and mountings.

Tools and Materials

Refrigerant line caps

Line wrenches

Graduated container

Protective Clothing

Goggles or safety glasses with side shields

Describe the vehicle being worked on:

Year ________ Make ____________________ Model ____________________

VIN ____________________ Engine type and size ____________________

PROCEDURE

1. Disconnect the negative battery cable. Task Completed ☐
2. Identify and disconnect all electrical connections to the compressor. Task Completed ☐
3. Discharge the system using a recovery/recycling machine. Task Completed ☐
4. Disconnect the refrigerant lines at the compressor. Task Completed ☐
5. Immediately cap or seal the ends of the lines or hoses. Task Completed ☐
6. Remove the drive belt for the compressor. Task Completed ☐
7. Loosen and remove the compressor mounting brackets. Note the location of the bolts, as they are typically a different length. Task Completed ☐
8. Remove the compressor. Task Completed ☐
9. Clean the outside of the compressor. Task Completed ☐

10. Drain the oil from the compressor into the graduated container. Allow at least ten minutes for the oil to drain. How much oil came out?

____

____

11. Using the service information, identify the type of oil and the amount that should be installed in the compressor.

____

____

12. Does the amount drained from the compressor equal the amount specified? If it is less, what is indicated?

____

____

13. Discard the old refrigerant oil according to the local and state requirements. What are these?

____

____

14. Add new oil in the correct amount as removed to the new or repaired compressor. Often, rebuilt and new compressors are pre-filled with the correct oil, therefore no oil should be added. Did you need to add oil to the compressor?

____

____

15. To install a compressor, reverse the removal procedure. Make sure the drive belt is tightened to the proper tension. Task Completed ☐

16. Recharge the system with refrigerant. Task Completed ☐

17. Does the air conditioning system now perform satisfactorily?

____

____

Problems Encountered

____

____

____

Instructor's Comments

____

____

____

HEATING AND AIR CONDITIONING JOB SHEET 21

Hybrid Vehicle A/C Systems

Name ______________________ Station ________________ Date ________________

NATEF Correlation

This Job Sheet addresses the following **MLR** task:

B.2. Identify hybrid vehicle A/C system electrical circuits, service, and safety precautions.

This Job Sheet addresses the following **AST/MAST** task:

B.4. Identify hybrid vehicle A/C system electrical circuits, service, and safety precautions.

Objective

Upon completion of this job sheet, you will be able to locate and safely disconnect the high-voltage circuit on a hybrid vehicle and safely work on and around the air conditioning system.

Tools and Materials

Service information

Vinyl tape for insulation

Protective Clothing

Goggles or safety glasses with side shields

Insulated gloves that are dry and are not cracked, ruptured, torn, or damaged in any way

Describe the vehicle being worked on:

Year __________ Make ________________________ Model ________________________

VIN ________________________ Engine type and size ________________________

PROCEDURE

WARNING: *Unprotected contact with any electrically charged ("hot" or "live") high-voltage component could cause serious injury or death.*

1. Describe how this vehicle is identified as a hybrid.

____

____

2. After referring to the vehicle's owner's manual, briefly describe its operation.

____

____

____

____

3. How many volts are the batteries rated at?

____

____

4. Where are the batteries located?

____

____

5. If the A/C compressor is powered by electricity, how many volts are used to operate the compressor?

____

____

6. How are the high-voltage cables labeled and identified?

____

____

7. What is used to provide short circuit protection in the high-voltage battery pack?

____

____

8. What isolates the high-voltage system from the rest of the vehicle when the vehicle is shut off?

____

____

9. Without touching the high-voltage cables or components, describe the routing of the cables. Include what they appear to be connected from and to.

____

____

____

____

WARNING: *Never assume that a hybrid vehicle is shut off simply because it is silent. Make sure the ignition key is in your pocket and not in the ignition switch.*

10. Describe the procedure for totally isolating the high-voltage system from the vehicle. This often involves the removal of a service plug. Be sure to include the location of this plug in your description of the procedure.

____

____

____

11. Make two signs saying, "WORKING ON HIGH-VOLTAGE PARTS. DO NOT TOUCH!" Attach one to the steering wheel, and set the other one near the parts you are working on. Why should you do this?

12. After the service plug has been removed, how long should you wait before working around or on the high-voltage system?

13. Describe the type of drive the A/C compressor uses.

14. Name the type of oil that must be installed in this compressor and explain why using the correct type is critical.

15. If the wrong type of oil has been installed in the compressor or A/C system, what must be done?

16. Refer to the service information and list at least five precautions that must be followed when servicing this A/C system.

17. To inspect or replace the compressor, the high-voltage system of the vehicle must be isolated. How is this done?

CAUTION: *Because the compressor has a high-voltage circuit, wear insulated gloves and isolate the high-voltage circuit before inspection or service.*

18. To remove the compressor, disconnect the power cables to the compressor and wrap the terminal ends with insulating tape before removing it from its mount. What color are the power cables?

    ____

19. After service, reconnect the battery and the high-voltage circuit. Some systems may need to be reinitialized, what are they?

    ____

    ____

    ____

Problems Encountered

____

____

____

Instructor's Comments

____

____

____

HEATING AND AIR CONDITIONING JOB SHEET 22

In-Line Filter Installation

Name ______________________ Station ______________ Date ______________

NATEF Correlation

This Job Sheet addresses the following **AST/MAST** task:

B.5. Determine the need for an additional A/C system filter; perform necessary action.

Objective

Upon completion of this job sheet, you will be able to determine whether or not the A/C system needs a new filter and be able to correctly install one if necessary.

Tools and Materials

In-line filter assembly

Line wrenches

Service information

Protective Clothing

Goggles or safety glasses with side shields

Describe the vehicle being worked on:

Year ________ Make ____________________ Model ____________________

VIN ____________________ Engine type and size ____________________

PROCEDURE

To determine if the system needs an in-line filter, answer the following questions:

1. Did the compressor need to be replaced because it was seized?

   ____

2. Does the system have a history of plugged metering devices?

   ____

3. If the answer to either of these questions was yes, the system needs to have a filter installed. Why would these conditions warrant a supplemental filter?

   ____

   ____

   ____

4. Describe the in-line filter. Does it have an orifice or is it just a filter?

____

____

____

5. If the filter does not have an orifice, locate a convenient location for installation in the liquid line between the condenser outlet and the evaporator inlet. Describe where you will install the filter.

____

____

____

6. If the filter has an orifice, find a convenient location for installation in the low-pressure side of the system beyond the expansion tube. Describe where you will install the filter.

____

____

____

7. Make sure the system is void of all refrigerant. How did you do this?

____

____

____

8. Secure the necessary tools, and then prepare the filter for installation according to the instructions that accompany the filter. What did you need to do?

____

____

____

9. Install the filter into the appropriate line. Task Completed ☐

10. If the filter had a built-in orifice, remove the original expansion tube from the system. Task Completed ☐

11. Make sure the lines are properly sealed, and then prepare to charge the system. What did you need to do before charging the system?

____

____

____

____

Problems Encountered

Instructor's Comments

HEATING AND AIR CONDITIONING JOB SHEET 23

R&R A/C System Components and Hoses

Name ______________________ Station ______________ Date ______________

NATEF Correlation

This Job Sheet addresses the following **AST/MAST** task:

B.6. Remove and inspect A/C system mufflers, hoses, lines, fittings, O-rings seals, and service valves; perform necessary action.

B.8. Remove, inspect, and reinstall receiver/drier or accumulator/drier; determine recommended oil quantity.

B.9. Remove, inspect, and install expansion valve or orifice (expansion) tube.

Objective

Upon completion of this job sheet, you will be able to remove and install various A/C components, including hoses and lines.

Tools and Materials

Manifold gauge set
Refrigerant recovery equipment
Vacuum pump
Orifice tube remover/installer
Hacksaw
Flare nut wrenches
Razor blades
Spring-lock tool

Protective Clothing

Goggles or safety glasses with side shields

Describe the vehicle being worked on:

Year ________ Make ____________________ Model ____________________

VIN ____________________ Engine type and size ____________________

PROCEDURE

1. Based on your diagnosis of the system, what components should be replaced?

____

____

____

2. Purge and recover the refrigerant in the system. Task Completed ☐
3. Disconnect the negative cable of the battery. Task Completed ☐

Refrigerant Oil

1. Whenever you replace a component, drain the oil from the component (after the component has been removed) into a graduated container. Record the amount of oil drained from each of the parts in this job sheet.

   a. Hoses

   ____

   b. Receiver/Drier

   ____

   c. Accumulator

   ____

   d. Thermostatic Expansion Valve

   ____

   e. Orifice Tube

   ____

2. What type of oil is specified for this A/C system?

   ____

3. According to the service information, how much oil should be added to the system when each of the following components is replaced?

   a. Hoses

   ____

   b. Receiver/Drier

   ____

   c. Accumulator

   ____

   d. Thermostatic Expansion Valve

   ____

   e. Orifice Tube

   ____

Hoses and Fittings

1. Although total hose replacement is the preferred way to correct for a hose leak, there are several accepted ways to repair refrigerant hoses and fittings. Using insert barb fittings and a length of replacement hose, you can fabricate an acceptable replacement for an original equipment hose. Insert barb fittings can also be used to replace the damaged section of a hose. The process for doing this is the same as what follows. To do this, measure and mark the required length of replacement high-pressure hose. Task Completed ☐

2. Using the razor blade, cut the hose to the desired length. Task Completed ☐
3. Examine the cut end and trim as necessary to have a square and flat cut. Task Completed ☐
4. Apply clean refrigerant oil to the inside of the hose. Task Completed ☐
5. Coat the fitting's insert with refrigerant oil and insert it into the hose. Task Completed ☐
6. Install the hose clamp and position it so it is over the barb closest to the fitting. Then, tighten it. Task Completed ☐
7. When using an insert barb fitting to replace a bad original equipment fitting, the original fitting must be removed without damaging too much of the hose. Remove the hose from the vehicle. Task Completed ☐
8. Set the hose in a vise to hold it while cutting through the fitting's ferrule with a hacksaw. Make this cut in the direction of the hose, not through the hose. Task Completed ☐
9. With the ferrule cut, use pliers to peel or pull the ferrule off. Task Completed ☐
10. Use the razor blade to cut the hose at a point just beyond the insert of the original fitting. Task Completed ☐
11. Examine the cut end and trim as necessary to have a square and flat cut. Task Completed ☐
12. Apply clean refrigerant oil to the inside of the hose. Task Completed ☐
13. Coat the fitting's insert with refrigerant oil and insert it into the hose. Task Completed ☐
14. Install the hose clamp and position it so that it is over the barb closest to the fitting. Then, tighten it. Task Completed ☐
15. If the hose or line has spring-lock fittings, a special tool is required to separate the sections. Put the tool over the coupling. Task Completed ☐
16. Close the tool and push the lines into the tool to release the female fitting from the garter spring of the coupling. Task Completed ☐
17. Pull the male and female fittings apart. Task Completed ☐
18. Remove the tool. Task Completed ☐
19. When joining two fittings with a spring-lock coupling, lubricate new O-rings and put them in their proper location on the male fitting. Task Completed ☐
20. Insert the male fitting into the female fitting. Task Completed ☐
21. Firmly push them together until they are secured by the spring-lock coupling. Task Completed ☐

Receiver/Drier

1. If the unit has a low-pressure switch, disconnect the wires to it. Task Completed ☐
2. Remove the inlet and outlet hoses at the receiver/drier. Task Completed ☐
3. Disconnect the mounting bolts and brackets from the receiver/drier. Task Completed ☐
4. Lift the receiver/drier out. Task Completed ☐
5. The procedure for installation is the reverse of removal. Task Completed ☐

Accumulator

1. Disconnect the lines at the accumulator's inlet and outlet fittings. Task Completed ☐
2. Loosen and/or remove the accumulator mounting bolts or screws. Task Completed ☐
3. Lift the accumulator out. Task Completed ☐
4. The procedure for installation is the reverse of removal. Task Completed ☐

Thermostatic Expansion Valve (TXV)

1. Disconnect the inlet and outlet lines at the TXV. Task Completed ☐
2. Remove whatever is used to keep the remote sensing bulb secure. Task Completed ☐
3. Loosen and/or remove the mounting clamp for the TXV. Task Completed ☐
4. Remove the TXV from the evaporator. Task Completed ☐
5. The procedure for installation is the reverse of removal. Task Completed ☐

Orifice Tube

1. Disconnect the inlet line at the evaporator. Task Completed ☐
2. Pour a small amount of clean refrigerant oil into the orifice tube. Task Completed ☐
3. Insert the orifice removal tool. Task Completed ☐
4. Turn the handle of the tool just enough for it to engage onto the tabs of the orifice tube. Task Completed ☐
5. Hold the handle of the tool in position while turning the tool's outer sleeve clockwise to remove the orifice tube. Task Completed ☐
6. Coat the new orifice tube with clean refrigerant oil. Task Completed ☐
7. Place the orifice tube into the evaporator line inlet. Push it in until it stops. Task Completed ☐
8. Install a new O-ring on the refrigerant line and reconnect it. Task Completed ☐

Problems Encountered

____

____

____

Instructor's Comments

____

____

____

HEATING AND AIR CONDITIONING JOB SHEET 24

Checking the Condenser for Airflow Restrictions

Name ______________________ Station ______________ Date ______________

NATEF Correlation

This Job Sheet addresses the following **MLR** task:

B.3. Inspect A/C condenser for airflow restrictions; determine necessary action.

This Job Sheet addresses the following **AST/MAST** task:

B.7. Inspect A/C condenser for airflow restrictions; determine necessary action.

Objective

Upon completion of this job sheet, you will be able to check the A/C condenser for airflow restrictions.

Tools and Materials

Flashlight

Service Information

Protective Clothing

Goggles or safety glasses with side shields

Describe the vehicle being worked on:

Year ________ Make ____________________ Model ____________________

VIN ____________________ Engine type and size ____________________

PROCEDURE

1. Describe the customer's concern.

____

____

____

2. Describe the problems that limited airflow through the condenser would cause.

____

____

____

3. Look over the condenser carefully, including the area between the radiator and condenser for any build-up of debris that might limit airflow.

____

____

____

4. Describe the effect that an inoperative cooling fan might have on the A/C system?

____

____

____

5. Summarize your findings on the vehicle.

____

____

____

Problems Encountered

____

____

____

Instructor's Comments

____

____

____

HEATING AND AIR CONDITIONING JOB SHEET 25

Servicing the Evaporator Water Drain

Name ______________________ Station ______________ Date ______________

NATEF Correlation

This Job Sheet addresses the following **AST/MAST** task:

B.10. Inspect evaporator housing water drain; perform necessary action.

Objective

Upon completion of this job sheet, you will be able to inspect the evaporator housing water drain.

Tools and Materials

Drain pan

Clean rags

Air nozzle

Protective Clothing

Goggles or safety glasses with side shields

Describe the vehicle being worked on:

Year ________ Make ________________ Model ________________

VIN ________________ Engine type and size ________________

PROCEDURE

1. Raise the vehicle on a lift. Task Completed ☐
2. Locate the evaporator case drain tube and describe its overall condition.

   ____

   ____

3. Place the drain pan under the evaporator case. Task Completed ☐
4. Disconnect the tube from the case. Did water come out of the case?

   ____

5. If no water came out, carefully clean out the drain hole and tube at the evaporator case. If it is necessary to insert a rod through the tube to clean it out, do this carefully because the tube is accordion-shaped and one of its bends may feel like a restriction. Task Completed ☐

6. If no water came out when you attempted to clear this tube, the entire evaporator case must be cleared. Describe your results so far.

____

____

7. If water came out of the case, the outside drain tube should be cleared by inserting the air nozzle into an end of the tube. Low air pressure should remove any restrictions in the tube. Describe what happened.

____

____

Problems Encountered

____

____

____

Instructor's Comments

____

____

____

HEATING AND AIR CONDITIONING JOB SHEET 26

Inspect, Test, and Service System Protection Devices

Name ______________________ Station ______________ Date ______________

NATEF Correlation

This Job Sheet addresses the following **MAST** task:

B.11. Diagnose A/C system conditions that cause the protection devices (pressure, thermal, and PCM) to interrupt system operation; determine necessary action.

Objective

Upon completion of this job sheet, you will be able to inspect and test system protection devices.

Tools and Materials

Fender covers	Test light
DMM	Service information

Protective Clothing

Goggles or safety glasses with side shields

Describe the vehicle being worked on:

Year ________ Make ____________________ Model ____________________

VIN ____________________ Engine type and size ____________________

Describe the type of air conditioning found on the vehicle:

PROCEDURE

1. Using the service information, locate the various pressure and temperature sensors in the system that controls the operation of the compressor. List them here along with a brief description of their purpose.

____

____

____

____

2. What abnormal operating conditions may cause the various switches or sensors to cause the A/C system not to operate normally?

____

____

____

3. Use the information to determine if each of these is normally closed or normally open. Label the sensors and switches with a NO or NC.

____

____

4. When checking the resistance of a NO switch, what should you measure across that switch when the switch is in normal conditions and is disconnected?

____

____

5. Check all NO switches with an ohmmeter and record your findings.

____

____

____

____

6. When checking the resistance of a NC switch, what should you measure across that switch when the switch is in normal conditions and is disconnected?

____

____

7. Check all NC switches with an ohmmeter and record your findings.

____

____

____

____

8. Summarize your determination about the condition of the switches and sensors.

____

____

9. Look in the service information to determine which of these sensors or switches can be replaced with refrigerant still in the system.

____

____

10. List any special precautions or procedures that must be followed when replacing the sensors and switches in this system.

____

____

Problems Encountered

Instructor's Comments

HEATING AND AIR CONDITIONING JOB SHEET 27

Removing and Installing an Evaporator

Name ______________________ Station ______________ Date ______________

NATEF Correlation

This Job Sheet addresses the following **AST** task:

B.11. Determine procedure to remove and reinstall evaporator; determine required oil quantity.

This Job Sheet addresses the following **MAST** task:

B.12. Determine procedure to remove and reinstall evaporator; determine required oil quantity.

Objective

Upon completion of this job sheet, you will be able to correctly remove and reinstall an evaporator. You will also be able to determine the amount of oil that should be added to the system after a new evaporator has been installed.

Tools and Materials

Charging station	Miscellaneous plugs and hoses
Gauge set	New refrigerant oil
Graduated container	Refrigerant recovery/recycling machine
Hand tools	Service information

Protective Clothing

Goggles or safety glasses with side shields

Describe the vehicle being worked on:

Year ________ Make ____________________ Model ____________________

VIN ____________________ Engine type and size ____________________

PROCEDURE

1. Recover the refrigerant using an approved recovery/recycling equipment. Task Completed ☐
2. Disconnect the negative battery cable. Task Completed ☐
3. If the evaporator and heater core are a combined unit, drain the engine's coolant. Task Completed ☐
4. Disconnect and label all electrical connectors, cables, and vacuum hoses that are connected to the evaporator. What did you need to disconnect?

   ____

   ____

5. Disconnect the refrigerant hoses at the evaporator and plug or cap the hose ends to prevent dirt and moisture from entering the system. Task Completed ☐

6. Unbolt and remove the evaporator. What did you need to remove to gain access to the evaporator?

____

____

7. Drain the oil from the evaporator into a graduated container. How much oil was drained from the evaporator?

____

____

8. Check the oil for dirt. If the oil is contaminated, replace the accumulator or receiver/drier. Describe the condition of the oil.

____

____

9. Add the preceding amount of new refrigerant oil or the amount specified in the service manual to the new evaporator. How much oil did you put into the evaporator?

____

____

10. Coat the new O-rings for the evaporator and the line fittings on the evaporator with clean refrigerant oil. Task Completed ☐

11. Install the new evaporator and tighten the fittings. Task Completed ☐

12. Reinstall and reconnect all parts, wires, cables, and vacuum hoses that were disconnected during removal. Add coolant if needed. Task Completed ☐

13. Connect the negative battery cable. Task Completed ☐

14. Evacuate and recharge the system. How much refrigerant was put in the system?

____

15. Perform a leak test and correct any problems. Were there any problems?

____

____

16. Drive the vehicle and then conduct a performance test. Does the system work as it should? Explain.

____

____

____

____

Problems Encountered

Instructor's Comments

HEATING AND AIR CONDITIONING JOB SHEET 28

Removing and Installing a Condenser

Name ______________________ Station ______________ Date ______________

NATEF Correlation

This Job Sheet addresses the following MAST task:

B.13. Remove, inspect, and reinstall condenser; determine required oil quantity.

Objective

Upon completion of this job sheet, you will be able to correctly remove and reinstall a condenser. You will also be able to determine the amount of oil that should be added to the system after a new condenser has been installed.

Tools and Materials

Miscellaneous plugs and hoses
New refrigerant oil
Refrigerant recovery/recycling machine
Gauge set
Graduated container
Service information

Protective Clothing

Goggles or safety glasses with side shields

Describe the vehicle being worked on:

Year ________ Make ____________________ Model ____________________

VIN ____________________ Engine type and size ____________________

PROCEDURE

1. Recover the refrigerant using an approved recovery/recycling equipment. Task Completed ☐
2. Disconnect the negative battery cable. Task Completed ☐
3. On many vehicles, the radiator, cooling fan, and/or shroud must be removed to gain access to the condenser. If the radiator must be removed, drain the coolant. What do you need to remove?

   ____

   ____

4. Disconnect and label all electrical connectors that are connected to the condenser and all those that may be in the way. What did you need to disconnect?

   ____

   ____

5. Disconnect the refrigerant hoses at the condenser and plug or cap the hose ends to prevent dirt and moisture from entering the system. Task Completed ☐

6. Unbolt and remove the evaporator. If the receiver/drier mounts to the condenser, disconnect it and remove it with the condenser. Task Completed ☐

7. Drain the oil from the condenser into a graduated container. How much oil was drained from the condenser?

___

8. Check the oil for dirt. If the oil is contaminated, replace the accumulator or receiver/drier. Describe the condition of the oil.

___

___

9. Add the preceding amount of new refrigerant oil or the amount specified in the service information to the new condenser. How much oil did you put into the evaporator?

___

___

10. Install the new condenser and loosely tighten the mounting bolts. Task Completed ☐

11. Coat the new O-rings for the condenser and the line fittings on the condenser with clean refrigerant oil. Task Completed ☐

12. Tighten the fittings, and then tighten the mounting bolts. Task Completed ☐

13. Reinstall and reconnect all parts, wires, cables, and vacuum hoses that were disconnected or removed during removal. Add coolant if needed. Task Completed ☐

14. Connect the negative battery cable. Task Completed ☐

15. Evacuate and recharge the system. How much refrigerant was put in the system?

___

16. Perform a leak test and correct any problems. Were there any problems?

___

___

17. Drive the vehicle and then conduct a performance test. Does the system work as it should? Explain.

___

___

___

Problems Encountered

Instructor's Comments

HEATING AND AIR CONDITIONING JOB SHEET 29

Checking Cooling System Hoses

Name ______________________ Station ______________ Date ______________

NATEF Correlation

This Job Sheet addresses the following **MLR/AST/MAST** task:

C.1. Inspect engine cooling and heater system hoses; perform necessary action.

Objective

Upon completion of this job sheet, you will be able to inspect and replace cooling and heater system hoses and inspect the drive belt for the A/C compressor.

Tools and Materials

Straightedge
Knife
Wire brush
Emery cloth
Belt tension gauge

Protective Clothing

Goggles or safety glasses with side shields

Describe the vehicle being worked on:

Year ________ Make ____________________ Model ____________________

VIN ____________________ Engine type and size ____________________

PROCEDURE

Hose Service

1. Carefully check all cooling hoses for leakage, swelling, and chafing. Record your findings here.

   ____

   ____

   ____

2. Squeeze each hose firmly. Do you notice any cracks or signs of splits when hoses are squeezed? Do any hoses feel mushy or extremely brittle?

   ____

   ____

   ____

3. Carefully examine the areas around the clamps. Are there any rust stains? What is or would be indicated by rust stains?

____

____

____

4. Carefully check the lower radiator hose. This hose contains a coiled wire lining to keep it from collapsing during operation. If the wire loses tension, the hose can partially collapse at high speed and restrict coolant flow. Describe this hose's condition.

____

____

5. Identify all hoses that need to be replaced.

____

____

____

____

____

6. Gather the new hoses, hose clamps, and coolant. Task Completed ☐

7. Release the pressure in the system by slowly and carefully opening the radiator cap. Task Completed ☐

8. Once the pressure is relieved, drain the coolant system below the level that is being worked on. Make sure the drained coolant is collected and recycled or disposed of according to local regulations. What did you do with the drained coolant?

____

____

____

____

9. Use a knife to cut off the old hose and loosen or cut the old clamp. Task Completed ☐

10. Slide the old hose off the fitting. If the hose is stuck, do not pry it off; instead, cut it off. Task Completed ☐

11. Clean off any remaining hose particles with a wire brush or emery cloth. What can happen if dirt or metal burrs are on the fitting when a new hose is attached?

____

____

12. Coat the surface with a sealing compound. What compound did you use?

____

13. Place the new clamps on each end of the hose before positioning the hose. Task Completed ☐
14. Slide the clamps to about 1/4 inch from the end of the hose after it is properly positioned on the fitting. Task Completed ☐
15. Tighten the clamp securely. Do not overtighten. Task Completed ☐
16. Refill the cooling system with coolant. Task Completed ☐
17. Run the engine until it is warm. Task Completed ☐
18. Bleed the system according to the manufacturer's recommendations. What did you do to bleed the system?

____

____

____

____

19. Turn off the engine and recheck the coolant level. Task Completed ☐
20. Retighten the heater and cooling system hose clamps. Task Completed ☐

Problems Encountered

____

____

____

Instructor's Comments

____

____

____

HEATING AND AIR CONDITIONING JOB SHEET 30

Testing Heater Control Valves

Name ______________________ Station ______________ Date ______________

NATEF Correlation

This Job Sheet addresses the following **AST/MAST** task:

C.2. Inspect and test heater control valve(s); perform necessary action.

Objective

Upon completion of this job sheet, you will be able to inspect and test heater control valves properly.

Tools and Materials

Hand-operated vacuum pump

Protective Clothing

Goggles or safety glasses with side shields

Describe the vehicle being worked on:

Year ________ Make ____________________ Model ____________________

VIN ____________________ Engine type and size ____________________

PROCEDURE

1. Locate the heater control valve and inspect the hoses' connections and their clamps. Summarize your findings.

   ____

   ____

2. If the valve is mechanically controlled, check the condition of the linkage and the cable. Summarize your findings.

   ____

   ____

3. If the valve is vacuum controlled, check the condition of the vacuum hose and connection at the valve. Summarize your findings.

   ____

   ____

4. Disconnect the vacuum hose and connect a hand-operated vacuum pump to the valve. Task Completed ☐

5. Pump a vacuum at the valve and watch its operation; it should open smoothly. Summarize your findings.

____

____

6. Refer to the service information to determine if the valve is normally open or closed. Turn the heater control on the instrument panel off. Check the valve to see if it is where it should be. Summarize your findings.

____

____

7. Move the heater control to full heat and check the position of the valve. Summarize your findings.

____

____

Problems Encountered

____

____

____

Instructor's Comments

____

____

____

HEATING AND AIR CONDITIONING JOB SHEET 31

Diagnosing Heater Temperature Control Problems

Name ______________________ Station ______________ Date ______________

NATEF Correlation

This Job Sheet addresses the following **MAST** task:

C.3. Diagnose temperature control problems in the heater/ventilation system; determine what caused the PCM to interrupt system operation; determine necessary action.

Objective

Upon completion of this job sheet, you will be able to diagnose temperature control problems in the heater/ventilation system.

Tools and Materials

Scan tool

Service Information

Protective Clothing

Goggles or safety glasses with side shields

Describe the vehicle being worked on:

Year ________ Make ____________________ Model ____________________

VIN ____________________ Engine type and size ____________________

PROCEDURE

Cooling System

1. When there is a problem of insufficient heat, begin your diagnosis with a visual inspection and a check of the coolant level. Summarize your findings.

 ____

 ____

2. Check the engine's cooling system for proper operation and signs of leaks. Summarize your findings.

 ____

 ____

3. If the cooling system seems to be fine, check the operation of the blower motor. Summarize your findings.

____

____

4. Turn the heater controls on and run the engine until it reaches normal operating temperature. Use the scan tool to determine the operating temperature of the engine. If the temperature is not within specifications, suspect a faulty thermostat. Summarize your findings.

____

____

5. If the engine is operating at the correct temperature, check the temperature of the two heater hoses. They should both be hot. If only one of the hoses is hot, suspect the heater control valve or a plugged heater core. Summarize your findings.

____

____

6. Most often, if these two items are faulty, the engine's cooling system will be negatively affected. Both of these items are replaced, rather than repaired. Summarize your findings.

____

____

Electronic Controls

1. Determine the control modules that control heating and air conditioning controls. List them here.

____

____

____

____

2. Using service information, determine the heating and air conditioning module sensors that can cause incorrect operation of the heating and ventilation systems. List them here.

____

____

____

3. Use the scan tool and list any DTCs that the vehicle has stored in memory for the heating and air conditioning. List them here.

____

____

____

Problems Encountered

____

____

____

Instructor's Comments

____

____

____

HEATING AND AIR CONDITIONING JOB SHEET 32

Removing and Installing a Heater Core

Name ______________________ Station ______________ Date ______________

NATEF Correlation

This Job Sheet addresses the following **AST** task:

C.3. Determine procedure to remove, inspect, and reinstall heater core.

This Job Sheet addresses the following **MAST** task:

C.4. Determine procedure to remove, inspect, and reinstall heater core.

Objective

Upon completion of this job sheet, you will be able to correctly remove and reinstall a heater core.

Tools and Materials

Catch basin — Coolant recycling machine

Hand tools — Service information

Protective Clothing

Goggles or safety glasses with side shields

Describe the vehicle being worked on:

Year ________ Make ____________________ Model ____________________

VIN ____________________ Engine type and size ____________________

PROCEDURE

1. Drain and recycle the coolant from the engine with the approved recycling equipment. Task Completed ☐
2. Disconnect the negative battery cable. Task Completed ☐
3. Refer to the service information for the recommended procedure for removing the heater core. Summarize the steps here.

   ____

   ____

   ____

   ____
4. If the evaporator and heater core are a combined unit, recover and recycle the refrigerant in the A/C system. Task Completed ☐

5. Disconnect and label all electrical connectors that are connected to the heater core and all those that may be in the way. What did you need to disconnect?

____

____

6. Loosen the hose clamps on the hoses connected to the heater core. Task Completed ☐

7. Remove the coolant hoses at the heater core and plug or cap the hose ends to prevent more loss of coolant. Task Completed ☐

8. Inspect the heater hoses and record your findings here.

____

____

9. Replace the heater hoses if necessary or desired. Task Completed ☐

10. Unbolt and remove the heater core. Task Completed ☐

11. Loosely place the new heater core into position. Task Completed ☐

12. Install new hose clamps over the hose ends. Task Completed ☐

13. Slip the hose ends into position over the inlet and outlet nipples on the heater core. Task Completed ☐

14. Tighten the hose clamps. Task Completed ☐

15. Tighten the mounting bolts for the heater core. Task Completed ☐

16. Reinstall and reconnect all parts, wires, cables, and vacuum hoses that were disconnected or removed during removal. Task Completed ☐

17. Fill the cooling system with new clean coolant and bleed the system. Task Completed ☐

18. Connect the negative battery cable. Task Completed ☐

19. Perform a leak test and correct any problems. Were there any problems?

____

____

20. Drive the vehicle and check the heating system. Does the system work as it should? Explain.

____

____

____

____

Problems Encountered

Instructor's Comments

HEATING AND AIR CONDITIONING JOB SHEET 33

Diagnosing Motor-Driven Accessories

Name ______________________ Station ______________ Date ______________

NATEF Correlation

This Job Sheet addresses the following **AST/MAST** task:

D.1. Inspect and test A/C-heater blower motors, resistors, switches, relays, wiring, and protection devices; perform necessary action.

Objective

Upon completion of this job sheet, you will be able to inspect and test A/C-heater blower motor circuits and components.

Tools and Materials

Wiring diagram for the vehicle

DMM

Protective Clothing

Goggles or safety glasses with side shields

Describe the vehicle being worked on:

Year ________ Make ____________________ Model ____________________

VIN ____________________ Engine type and size ____________________

PROCEDURE

NOTE: *Although this job sheet is focused on the blower motor, the procedures can easily be applied to all motor-driven accessories. Refer to the appropriate service manual and wiring diagram to identify the components of the circuit you wish to test. Then, apply the same sequence and logic given in this job sheet.*

1. Refer to the service information to identify the number of speeds at which the blower motor should operate. Also, identify whether the motor should run when the switch is in the off position or its lowest position. Describe how the blower motor should operate.

   ____

   ____

2. Check the wiring diagram and identify whether the blower circuit is controlled by a ground side switch or an insulated (power side) switch. State which one.

   ____

   ____

3. Start the engine and turn on the blower. Move the blower control to all available positions and summarize what happened in each.

____

____

4. Match the type of circuit control and the problem with one of the following.

The blower works at some speeds but not all.

a. Check the voltage to the blower motor at the various switch positions. Task Completed ☐

b. If the voltage doesn't change when a new position is selected, check the circuit from the switch to the resistor block. Task Completed ☐

c. If there was zero voltage in a switch position, check for an open in the resistor block. Task Completed ☐

d. Give a summary of the test results.

____

____

The blower doesn't work and is controlled by an insulated switch.

a. Check the fuse or circuit breaker. If it is open, check the circuit for a short. Task Completed ☐

b. Connect a jumper wire from a power source to the motor. If the motor operates, there is an open in the circuit between the fuse and the motor. Task Completed ☐

c. To identify the location of the open, connect a jumper wire across the switch. If the motor now operates, the switch is bad. Task Completed ☐

d. Connect a jumper wire across the resistor block. If the motor now operates, the resistor block is bad. Task Completed ☐

e. If the motor did not operate when the jumper wire connected it to a power source, connect the jumper wire from the motor to a known good ground. If the motor operates, the fault is in the ground circuit. If the motor doesn't operate, the motor is bad. Task Completed ☐

f. Give a summary of the test results.

____

____

The blower doesn't work and is controlled by a ground side switch.

a. Check the fuse or circuit breaker. If it is open, check the circuit for a short. Task Completed ☐

b. Connect a jumper wire from the motor's ground terminal to a known good ground. If the motor runs, the problem is in that circuit. Task Completed ☐

c. If the motor did not operate, check for voltage at the battery terminal of the motor. If no voltage is found, there is an open in the power feed circuit for the motor. If voltage was present, the motor is bad. Task Completed ☐

d. If the motor operated when the jumper wire was connected in step b, use an ohmmeter to check the ground connection of the switch. Task Completed ☐

e. If the ground is good, use a voltmeter to probe for voltage at any of the circuits from the resistor block to the switch. Replace the switch if there is power at that point. Task Completed ☐

f. If no voltage was available at the switch, check the circuit for an open. If there is not an open, suspect a faulty switch. Task Completed ☐

g. Give a summary of the test results.

___

___

The blower fan runs constantly.

a. If the motor runs when it should not, there is probably a short in the circuit. If a ground side switch controls the circuit, check for a short to ground in the control circuit. The exact problem can be isolated by disconnecting portions of the circuit until the motor stops. The short is in the part of the circuit that disconnected last. Task Completed ☐

b. If the circuit is controlled by an insulated switch, check for a wire-to-wire short. Check other circuits of the vehicle to identify what circuit is involved in this problem. That circuit will also experience a lack of control, or when that circuit is turned off, the blower motor will turn off. The exact problem can be isolated by disconnecting portions of the circuit until the motor stops. The short is in the part of the circuit that disconnected last. Task Completed ☐

c. Give a summary of the test results.

___

___

Problems Encountered

___

___

___

Instructor's Comments

___

___

___

HEATING AND AIR CONDITIONING JOB SHEET 34

Servicing Compressor Load Sensors

Name ______________________ Station ______________ Date ______________

NATEF Correlation

This Job Sheet addresses the following **AST/MAST** task:

D.2. Diagnose A/C compressor clutch control systems; determine necessary action.

Objective

Upon completion of this job sheet, you will be able to demonstrate the ability to accurately test A/C compressor load cutoff systems.

Tools and Materials

DMM

Service information

Protective Clothing

Goggles or safety glasses with side shields

Describe the vehicle being worked on:

Year ________ Make ____________________ Model ____________________

VIN ____________________ Engine type and size ____________________

PROCEDURE

1. Using the service information, locate the various pressure and temperature sensors in the system that control the operation of the compressor. List them and describe their location.

   ____

   ____

   ____

   ____

2. Check the information to determine if each of these is normally closed or normally open, and summarize your findings.

   ____

   ____

   ____

3. Checking one sensor at a time, complete the following steps.
   a. Disconnect the wires at the sensor or switch.
   b. Connect the DMM across the switch terminals.
   c. Set the meter for resistance or continuity checks.
   d. If the switch is normally closed, the reading should be zero ohms.
   e. If the switch is normally open, you should get an infinite reading.
   f. Any reading other than these is an indication of a faulty switch.
   g. Summarize your results from each switch.

____

____

____

____

____

4. If a switch needs to be replaced, the system may need to be discharged. Check the service information before proceeding. Which switches can be replaced without discharging the refrigerant in the system?

____

____

____

____

5. To remove the defective switch, loosen it and remove it. Task Completed ☐
6. To install the new switch, carefully thread it into place and tighten it. Task Completed ☐
7. Reconnect the wires leading to the switches. Task Completed ☐

Problems Encountered

____

____

____

Instructor's Comments

____

____

____

HEATING AND AIR CONDITIONING JOB SHEET 35

Testing HVAC Controls

Name ______________________ Station ______________ Date ______________

NATEF Correlation

This Job Sheet addresses the following **AST/MAST** tasks:

D.3. Diagnose malfunctions in the vacuum, mechanical, and electrical components and controls of the heating, ventilation, and A/C (HVAC) system; determine necessary action.

D.5. Inspect and test A/C-heater control cables, motors, and linkages; perform necessary action.

Objective

Upon completion of this job sheet, you will be able to diagnose failures in the electrical, vacuum, and mechanical controls of heating, ventilation, and air conditioning systems.

Tools and Materials

Service information

Vacuum gauge

Hand-operated vacuum pump

Protective Clothing

Goggles or safety glasses with side shields

Describe the vehicle being worked on:

Year ________ Make ____________________ Model ____________________

VIN ____________________ Engine type and size ____________________

PROCEDURE

1. Check the fuses for the compressor clutch and the blower motor. Summarize your findings.

   ____

   ____

2. Check the operation of the blower switch in all switch positions. Summarize your findings.

   ____

   ____

3. If there is a problem with the blower motor, check the circuit to determine the exact fault. Then, summarize your findings.

____

____

4. With the engine running, turn the A/C on and off and listen for the engagement of the compressor clutch. If the clutch makes a chattering noise, it may need adjustment or replacement. Summarize your findings.

____

____

5. If the clutch did not cycle with the switch, check to see if the clutch is constantly engaged or disengaged. If always engaged, the clutch is bad or there is a wire-to-wire short providing voltage to the clutch at all times. If the clutch is always disengaged, there is probably a problem with the clutch control circuit. Summarize your findings.

____

____

6. Check the electrical connections at the clutch. Summarize your findings.

____

____

7. Connect a jumper wire from the battery to the clutch. Check its operation now. If it operates quietly and smoothly, the problem is in the clutch control circuit. If it didn't operate normally when jumped, the clutch is bad. Summarize your findings.

____

____

8. Check the resistance of the clutch coil with an ohmmeter. Compare your measurement to specifications. Summarize your findings.

____

____

9. Using the service information, identify all sensors or switches that control or limit the operation of the clutch. List all of them and summarize their condition.

____

____

10. Identify and inspect the hoses and components for vacuum motors and doors. Look for disconnected or broken hoses, broken connectors, misrouted vacuum lines, and loose or disconnected electrical connectors at vacuum switches and controls. Summarize your findings.

____

____

11. Check the vacuum at the engine's intake manifold. Do this while the engine is idling. Summarize your findings.

____

____

12. Check the individual vacuum components. Use a hand-operated vacuum pump to check the component's activity and its ability to hold a vacuum. Summarize your findings.

____

____

13. Inspect all mechanical or cable linkages and controls. Make sure the cables are properly attached to the levers of the control switch. Summarize your findings.

____

____

14. If the cable(s) is equipped with an automatic adjuster, make sure the adjusting mechanism operates freely and is not damaged. Summarize your findings.

____

____

Problems Encountered

____

____

____

Instructor's Comments

____

____

____

HEATING AND AIR CONDITIONING JOB SHEET 36

Checking an A/C Master Control Unit

Name ______________________ Station ______________ Date ______________

NATEF Correlation

This Job Sheet addresses the following **AST/MAST** task:

D.4. Inspect and test A/C-heater control panel assembly; determine necessary action.

Objective

Upon completion of this job sheet, you will be able to inspect and test an A/C-heater control panel assembly.

Tools and Materials

DMM

Service information

Hand-operated vacuum pump

Protective Clothing

Goggles or safety glasses with side shields

Describe the vehicle being worked on:

Year ________ Make ____________________ Model ____________________

VIN ____________________ Engine type and size ____________________

PROCEDURE

1. Remove the A/C-heater master control from the instrument panel. Task Completed ☐
2. Inspect the connectors that go to the control unit and record your findings. Look for damage and corrosion.

   ____

   ____

3. Thoroughly inspect the connectors at the control unit. Look for terminal distortion (bends), corrosion, and other damage. Record your findings.

   ____

   ____

4. With an ohmmeter, check the continuity through the blower motor switch in all its positions. Summarize your findings.

   ____

   ____

5. Compare your results with the wiring diagram for the switch. Do you have good continuity at the correct terminals when the switch is in its various positions? ____________________

6. With the ohmmeter, check the resistance across the temperature control switch. Watch the meter as you sweep through the full range of the switch, and record your findings.

____

____

____

7. If the mode selector switch is electrical, check for continuity across the terminals as you move the switch through the various mode selections. Record your findings.

____

____

____

8. If the mode selector switch is a vacuum switch, apply a vacuum to the inlet of the switch and feel for a vacuum at the various tube connectors on the switch while you move the selector through its various positions. Record your findings.

____

____

____

9. Compare your results to the information in the service manual, and summarize your recommendations for service to the master control unit.

____

____

____

____

10. Reinstall the old switch or install a new one. Task Completed ☐

Problems Encountered

____

____

____

Instructor's Comments

____

____

____

HEATING AND AIR CONDITIONING JOB SHEET 37

Inspecting the Air Duct System

Name ______________________ Station ______________ Date ______________

NATEF Correlation

This Job Sheet addresses the following **MLR** task:

D.1. Inspect A/C-heater ducts, doors, hoses, cabin filters, and outlets; perform necessary action.

This Job Sheet addresses the following **AST/MAST** task:

D.6. Inspect A/C-heater ducts, doors, hoses, cabin filters, and outlets; perform necessary action.

Objective

Upon completion of this job sheet, you will be able to inspect and test the A/C-heater ducts, doors, hoses, and outlets.

Tools and Materials

Service information

Scan tool

Hand tools

Protective Clothing

Goggles or safety glasses with side shields

Describe the vehicle being worked on:

Year __________ Make ____________________ Model ____________________

VIN ____________________ Engine type and size ____________________

PROCEDURE

1. Using service information, describe the operation of the HVAC mode doors.

____

____

2. Check the HVAC system for trouble codes, if applicable. Summarize your findings.

____

____

3. Test the operation of the HVAC system in each of the following settings and describe your results.

a. MAX ____________________

b. NORM ____________________

c. Bilevel ______________________

d. VENT ______________________

e. HEAT ______________________

f. BLEND ______________________

g. DEFROST ______________________

h. OFF ______________________

4. Compare your findings to the specifications given in the service information, and state your conclusions and recommendations for system service.

____

____

____

____

5. The in-cabin dust and pollen filter should be replaced at a fixed interval, normally every 30,000 miles (48,000 km) or 24 months. When is it recommended that the filter be changed on this vehicle?

____

____

6. With the blower on high speed, feel the airflow from the ducts and vents. If the flow seems to be less than it should, replace the filter. What did you find?

____

____

7. Normally, to replace the filter the glove box needs to be removed to access the filter's holder. Does the glove box need to be removed on this vehicle?

____

____

Problems Encountered

____

____

____

Instructor's Comments

____

____

____

HEATING AND AIR CONDITIONING JOB SHEET 38

Identifying the Source of Odors

Name ______________________ Station ______________ Date ______________

NATEF Correlation

This Job Sheet addresses the following **MLR** task:

D.2. Identify the source of A/C system odors.

This Job Sheet addresses the following **AST/MAST** task:

D.7. Identify the source of A/C system odors.

Objective

Upon completion of this job sheet, you will be able to identify the source of unwanted interior odors entering the vehicle.

Tools and Materials

Service information

Protective Clothing

Goggles or safety glasses with side shields

Describe the vehicle being worked on:

Year ________ Make ____________________ Model ____________________

VIN ____________________ Engine type and size ____________________

PROCEDURE

1. A common customer concern is a moldy and musty smell coming from the A/C system. What is the most likely cause of this?

____

____

2. Why does moisture collect on the evaporator of a good working A/C system?

____

____

3. If dampness is found on the carpet around the evaporator and there is a moldy odor, what is the most likely cause?

____

____

4. What is the purpose of the cabin air filter?

____

____

5. How often should the cabin air filter be replaced in this vehicle?

____

____

6. Where is the cabin air filter on this vehicle?

____

____

7. What other things could cause unwanted odors inside the vehicle?

____

____

Problems Encountered

____

____

____

Instructor's Comments

____

____

____

HEATING AND AIR CONDITIONING JOB SHEET 39

Checking Automatic Climate Control Systems

Name ______________________ Station ______________ Date ______________

NATEF Correlation

This Job Sheet addresses the following **AST/MAST** task:

D.8. Check operation of automatic or semi-automatic heating, ventilation, and air conditioning (HVAC) control systems; determine necessary action.

Objective

Upon completion of this job sheet, you will be able to check the operation of automatic and semiautomatic HVAC control systems.

Tools and Materials

Scan tool

DMM

Service information

Protective Clothing

Goggles or safety glasses with side shields

Describe the vehicle being worked on:

Year ________ Make ____________________ Model ____________________

VIN ____________________ Engine type and size ____________________

PROCEDURE

1. Refer to the service information and determine if the automatic climate control system relies on the PCM, body control module (BCM), or a separate computer for control. What did you find?

____

____

2. If the unit is part of the PCM or BCM system, connect the scan tool and retrieve any codes that may be present. What did you find?

____

____

3. If the unit is controlled by its own computer system, use the service information and record the procedure for retrieving trouble codes.

____

____

____

4. If there were trouble codes, what problem areas were identified?

____

____

5. Test the components or subsystems identified by the trouble codes, and summarize the results of those tests.

____

____

6. If the system is a semiautomatic system, check the service information for the proper diagnostic procedures for checking evaporator and heating controls. Record these procedures here.

____

____

____

7. Follow the appropriate test procedures and summarize your findings.

____

____

____

____

____

Problems Encountered

____

____

____

Instructor's Comments

____

____

____

HEATING AND AIR CONDITIONING JOB SHEET 40

Checking the Operation of Refrigerant Handling Equipment

Name ______________________ Station ______________ Date ______________

NATEF Correlation

This Job Sheet addresses the following **AST/MAST** task:

E.1. Perform correct use and maintenance of refrigerant handling equipment according to equipment manufacturer's standards.

Objective

Upon completion of this job sheet, you will be able to verify correct operation and maintenance of refrigerant handling equipment.

Tools and Materials

Manifold gauge sets

Refrigerant storage containers

Refrigerant recovery machines

Protective Clothing

Goggles or safety glasses with side shields

PROCEDURE

1. Check the manifold gauge set to make sure it has a cut off valve at the end of each hose so that the fitting not in use is automatically shut. Summarize your findings.

   ____

2. Check to make sure the manifold gauge set is clearly labeled as to what refrigerant it should be used for. Manifold gauge sets for R-134a can be identified by one or all of the following: Labeled FOR USE WITH R-134a; labeled HFC-134 or R-134a; and/or have a light blue color on the face of the gauges. Summarize your findings.

   ____

3. Verify that the R-134a service hoses have a black stripe along their length and are clearly labeled SAE J2196/R-134a. The low pressure hose is blue with a black stripe, the high pressure hose is red with a black stripe, and the center service hose is yellow with a black stripe. Summarize your findings.

____

____

____

____

____

4. Make sure the containers of refrigerant are clearly labeled. When you are not sure of the refrigerant stored in a container or if you suspect a mixing of refrigerants has occurred, you should run a purity test and/or use a refrigerant identifier. Summarize your findings.

____

____

5. Check the refrigerant storage containers. All recycled refrigerant must be safely stored in DOT CFR Title 49 or UL-approved containers. Containers specifically made for R-134a should be so marked. Before any container of recycled refrigerant can be used, it must be checked for non-condensable gases. Summarize your findings.

____

____

6. Locate the UL approved label on the refrigerant recovery machines. Summarize your findings.

____

____

Problems Encountered

____

____

____

Instructor's Comments

____

____

____

HEATING AND AIR CONDITIONING JOB SHEET 41

Identify and Recover Refrigerant

Name ______________________ Station ______________ Date ______________

NATEF Correlation

This Job Sheet addresses the following **AST/MAST** task:

E.2. Identify and recover A/C system refrigerant.

Objective

Upon completion of this job sheet, you will be able to identify and recover refrigerant.

Tools and Materials

Refrigerant identifier — Recovery/recycling machine

Manifold gauge set — Service information

Protective Clothing

Goggles or safety glasses with side shields

Describe the vehicle being worked on:

Year __________ Make ____________________ Model ____________________

VIN ____________________ Engine type and size ____________________

Describe the type of air conditioning found on the vehicle:

____

PROCEDURE

1. Check the decal under the hood on the air conditioning unit for the type of refrigerant used in your specific system. What did you find?

   ____

2. If this decal is not present, refer to the service information for your specific vehicle. Identify the type of refrigerant that should be used in this system. Record your findings.

   ____

NOTE: *It is very important to determine the type of refrigerant in your system before servicing. Different types of refrigerant cannot be mixed.*

3. Before proceeding, it is important to identify the refrigerant currently in the system. This includes any contaminants and refrigerant blends. Why is this important?

____

____

4. Connect the refrigerant identifier to the system. How did you do this?

____

____

____

5. Record the results of the test.

____

____

6. What do the test results tell you about recycling the current refrigerant?

____

____

7. Connect the manifold gauge set to the vehicle with the center hose of the set to the recovery/recycling unit. Describe what you need to do in order to recycle/recover the refrigerant.

____

____

8. When the recovery is complete, close the valves on the manifold gauge set and turn off the recovery unit. You are now prepared to do necessary repairs on the air-conditioning system. What service will be performed?

____

____

Problems Encountered

____

____

____

Instructor's Comments

____

____

____

HEATING AND AIR CONDITIONING JOB SHEET 42

Recycle, Label, and Store Refrigerant

Name ______________________ Station ______________ Date ______________

NATEF Correlation

This Job Sheet addresses the following **AST/MAST** task:

E.3. Recycle, label, and store refrigerant.

Objective

Upon completion of this job sheet, you will be able to recycle, label, and store refrigerant safely and properly. You will also be able to test recycled refrigerant for non-condensable gases according to the guidelines set by the SAE.

Tools and Materials

Recycling machine

Service information

Pressure gauge set

Thermometer

SAE pressure/temperature chart

Protective Clothing

Goggles or safety glasses with side shields

Describe the vehicle being worked on:

Year ________ Make ____________________ Model ____________________

VIN ____________________ Engine type and size ____________________

Describe the type of air conditioning found on the vehicle:

____

PROCEDURE

1. Refer to the manual for your specific machine to determine the procedure to recycle refrigerant. Briefly explain that procedure.

____

____

____

____

2. Use a refrigerant identifier and describe the contents of the refrigerant that is being removed from the vehicle. What were the results?

___

___

___

3. Collect the old refrigerant into suitable containers. Label the container as "empty" or "empty, evacuated and ready for disposal" or "Freon," "contains recycled Freon," "contains R-134a," "contains recycled R-134a," or "contains contaminated refrigerant." What did you need to label the containers as?

___

___

4. After the container has been identified, it can be stored or disposed of accordingly. However, the container must be stored at a temperature of 65°F (18.3°C) or above for 12 hours and protected from the sun. Where did you put the container?

___

___

5. Prior to using stored, recycled refrigerant, install a calibrated pressure gauge onto the container and measure the pressure in the container. What pressure was read?

___

6. Measure the temperature of the air 4 inches away from the container's surface. What was your measurement?

___

7. Compare the measured pressure and temperature to the pressure/temperature chart. Determine if the recycled refrigerant has excessive non-condensable gases. Explain your findings.

___

___

8. If the refrigerant has excessive non-condensable gases, it should be recycled again and then retested before it is used. Summarize the results of this test.

___

___

___

Problems Encountered

___

___

___

Instructor's Comments

___

___

___

HEATING AND AIR CONDITIONING JOB SHEET 43

Evacuating and Charging an A/C System

Name ______________________ Station ______________ Date ______________

NATEF Correlation

This Job Sheet addresses the following **AST/MAST** task:

E.4. Evacuate and charge A/C system; add refrigerant oil as required.

Objective

Upon completion of this job sheet, you will be able to evacuate and recharge an A/C system with a manifold gauge set and a recycling and charging station.

Tools and Materials

Manifold gauge set

Refrigerant recycling/reclaiming machine

Vacuum pump

Proper adapters for the fittings

Protective Clothing

Goggles or safety glasses with side shields

Describe the vehicle being worked on:

Year ________ Make ____________________ Model ____________________

VIN ____________________ Engine type and size ____________________

Type of refrigerant used in the system: ______________

PROCEDURE

Using a Manifold Gauge Set

1. Make sure the refrigerant has been removed from the system with a refrigerant reclaiming machine. Do not vent the refrigerant into the atmosphere. Task Completed ☐
2. Close the hand valves on the gauge set. Task Completed ☐
3. Connect the hoses for manifold gauge set to the high-side and low-side service ports. Task Completed ☐
4. If the system is equipped with shutoff type service valves, open them slightly. Task Completed ☐
5. Connect the center manifold service hose to the intake of the vacuum pump. Task Completed ☐
6. Open the shutoff valves at each of the three service hoses. Task Completed ☐

7. Turn on the vacuum pump. Task Completed ☐
8. Open the low-side manifold hand valve and watch the low-side gauge. Did it drop to indicate a slight vacuum?

____

____

9. Once the gauge reads about 25 in. Hg, close the low-side hand valve and watch the gauge reading. Is the system holding the vacuum? If not, what is indicated?

____

____

10. Reopen the low-side hand valve. Task Completed ☐
11. Open the high-side hand valve. Task Completed ☐
12. Allow the vacuum pump to run until close to 30 in. Hg is showing on the low-side gauge. About how long did this take?

____

____

13. Close both hand valves. Task Completed ☐
14. Refer to the service information and identify the proper type and quantity of refrigerant to use in this system and record this here.

____

____

15. Will you be using pound cans or a larger can of refrigerant to charge the system?

____

16. Make sure the shutoff valves, manifold hand valves, and compressor service valves are closed. Task Completed ☐
17. Connect the center manifold service hose to the container of refrigerant. Task Completed ☐
18. Open the valve on the container. Task Completed ☐
19. Open the service hose, high-side, and low-side shutoff valves. Task Completed ☐
20. Start the engine and run it at a fast idle. Task Completed ☐
21. Turn on the A/C and set it and its blower to its maximum position. Task Completed ☐
22. Open the low-side manifold hand valve. Task Completed ☐
23. Allow only the recommended amount of refrigerant to enter the system. Task Completed ☐
24. Once the system is fully charged, close the low-side manifold hand valve. Task Completed ☐
25. Close the service hose shutoff valve. Task Completed ☐

26. Disconnect the service hose from the container. Task Completed ☐

27. Observe the pressures on the gauges. What do they indicate?

____

____

28. Return the engine to a normal idle speed and turn the A/C system off. Task Completed ☐

29. Turn off the engine. Task Completed ☐

30. Close all valves and use a recovery system to remove any refrigerant from the manifold and its hoses. Task Completed ☐

31. Remove the manifold gauge set. Task Completed ☐

32. Replace all protective caps and covers. Task Completed ☐

Using a Recycling and Charging Station

1. Physically locate the pressure fittings of the system you will be working with. Describe their location.

____

____

2. Make sure you have the proper adapters for the fittings prior to starting any service work on air conditioning systems. Task Completed ☐

3. Connect a gauge set to the system. Task Completed ☐

4. Connect the reclaiming unit to the gauge set. Task Completed ☐

5. Activate the reclaimer to discharge the system, and then begin evacuation. Allow enough time for the machine to draw a good vacuum (at least 29 in. Hg.) in the system. Observe the gauge set to determine when a vacuum is created. How long did it take to pump down the system?

____

6. Stop evacuation and let the system remain in a vacuum state for at least ten minutes. Check the vacuum at that time. Did the vacuum remain in place? What does this indicate?

____

____

7. Close the valves at the gauge set and disconnect the reclaiming machine. Task Completed ☐

8. Connect the charging station and open the gauge set valves. Task Completed ☐

9. Set the unit to deliver the required amount of refrigerant for that system. What is the proper charge?

____

10. After the system has been recharged, check the gauge readings while the system is in operation to make sure it is working properly. The gauge readings are:

Problems Encountered

Instructor's Comments